BEING THERE

Anthropology, Culture and Society

Series Editors:
Professor Thomas Hylland Eriksen, University of Oslo
Dr Jon P. Mitchell, University of Sussex

BEING THERE
FIELDWORK IN ANTHROPOLOGY

EDITED BY
C. W. WATSON

Pluto Press
LONDON ANN ARBOR, MI

First published 1999 by Pluto Press
345 Archway Road, London N6 5AA

Copyright © C. W. Watson 1999

British Library Cataloguing in Publication Data
A catalogue record for this book is available from
the British Library
ISBN 9780745314921 pbk
ISBN 0745314929 pbk
ISBN 9780745314976 hbk
ISBN 074531497X hbk

Library of Congress Cataloging in Publication Data
Being there : six anthropological accounts of fieldwork / edited by
C. W. Watson.
    p.   cm. — (Anthropology, culture, and society)
Includes bibliographical references.
ISBN 0–7453–1497–X (hbk)
    1. Anthropology—Field work. 2. Anthropology—Philosophy.
I. Watson, C. W. II. Series.
GN34.3.F53B45   1999
301'.07'23—dc21   98–37448
                                                            CIP

Designed and produced for Pluto Press by
Chase Production Services, Chadlington, OX7 3LN
Typeset from disk by Stanford DTP Services, Northampton
Printed on Demand by Antony Rowe Ltd

CONTENTS

NOTES ON CONTRIBUTORS

Andrew Beatty is attached to Wolfson College, Oxford. He has conducted extensive fieldwork in two different regions in Indonesia and is the author of *Society and Exchange in Nias* (1992) and *Varieties of Javanese Religion: An Anthropological Account* (in press).

Katy Gardner lectures in the School of African and Asian Studies at the University of Sussex. She is the author of *Songs from the River's Edge* (1991), a set of autobiographical short stories based on her fieldwork experience in Sylhet, Bangladesh, and a monograph entitled *Global Migrants: local lives, travel and transformation in rural Bangladesh* (1995). She is also the co-author (with David Lewis) of *Anthropology, Development and the Post-modern Challenge* (Pluto Press, 1996). She is currently carrying out research on Bangladesh Sylheti migrants in Britain.

Anna Grimshaw is a Lecturer in Visual Anthropology at the Granada Centre for Visual Anthropology, University of Manchester. The fieldwork for her PhD thesis was carried out in northern India. She is the author of *Servants of the Buddha* (1992), an account of her experience living with a community of nuns in the Himalayas.

Allison James teaches in the Department of Social Anthropology in the University of Hull. She has done extensive anthropological research in Britain. She is the author of *Childhood Identities* (1993), and (with C. Jenks and A. Prout) *Theorizing Childhood* (1990). She is currently engaged in a study of the social organisation of children's time.

Cris Shore is a lecturer in the Department of Anthropology at Goldsmith's College, University of London. He has worked extensively in Europe, and his research for his doctorate was carried out in Perugia.

He is the author of *Italian Communism: the Escape from Leninism. An Anthropological Approach* (1990), and (with Sue Wright) has edited *Anthropology of Policy: Critical Perspectives on Government and Power* (1997). He is currently writing a book on the anthropology of the European Union based on fieldwork among EU civil servants in Brussels.

C. W. (Bill) Watson teaches in the Department of Anthropology at the University of Kent at Canterbury. He has been doing fieldwork in Indonesia for a number of years. He has edited (with Roy F. Ellen) *Understanding Witchcraft and Sorcery in Southeast Asia* (1994) and is the author of *Kinship, Property and Inheritance in Kerinci, Central Sumatra* (1994).

INTRODUCTION: THE QUALITY OF BEING THERE

C. W. Watson

If, very broadly speaking, we think of anthropology as the discipline of explaining the behaviour and thoughts of people bounded within a culture in terms that are intelligible to people outside that culture, then fieldwork is that part of the process which takes place when the anthropologist is in the field dwelling among the people she hopes to describe. This is not to say that the anthropologist only begins to acquire knowledge once she is in the field; we are all aware how significant the pre-fieldwork period is for establishing concepts, understandings and notions of the people and culture we shall be encountering. Even when, as often happens, experience in the field leads us to react against our own earlier views, the tenacity of our preconceived ideas through the very process of reaction contributes substantially to how we formulate our knowledge of the time. Nor, on the other hand, is the emphasis here on fieldwork meant to suggest that the period of post-fieldwork, when one has left the field and is engaged in constructing the ethnography, is any less significant in the process of coming to an understanding. Again, as we generally acknowledge, reflection at a temporal and spatial distance from our experience within a different cognitive and experiential context inevitably brings about further reformulations and recastings of our thoughts and ideas and the best way of expressing them. There are yet further stages of anthropological creativity, perhaps not so commonly recognised, even after the ethnography has been secured in a text. Anthropologists – with increasing frequency these days as a consequence of ever-improving communications – often return to the field, pick up the threads and find themselves weaving a very differently patterned cloth from that which they wove so confidently during their previous encounter (for

1

example, Caplan 1992). A variation of this return to the field occurs when anthropologists revisit their own writings and find themselves unhappy with a description, irritated at the lacunae and generally dissatisfied with the inadequacies of their presentation. Before, during, after and second-thoughts – all these, then, are different stages that the anthropologist passes through in the never-to-be-completed task of full understanding and perfect translation.

It may therefore seem deceptive or deliberately misleading to mark off the period in the field as being worthy of special consideration when giving an account of the way anthropological knowledge is created. But while taking this point, and acknowledging that the period in the field is simply part of the ongoing temporal experience of coming to an understanding of other people, most anthropologists would, I think, recognise that although in terms of intentionality and intellectual endeavour there is no clear break in the temporal flux, the sheer physical, emotional and psychological experience of being in the field gives that time a unique quality, one which indelibly impresses itself on the personality and self-understanding of an anthropologist: things are never quite the same again.

This, then, might be one justification for giving fieldwork such prominence in any description of anthropology: it is a period of particularly heightened intensity. More conventionally, however, the justification would be not in drawing attention to the personal significance of fieldwork for the anthropologist – though there is still a powerful lobby within the profession that argues, very much like psychoanalysts for psychoanalysis, that one cannot be an anthropologist without having undergone that *rite de passage* which is constituted by fieldwork[1] – but in arguing that the period in the field is one of intensive, close engagement of a limited duration, in which through the employment of various techniques and strategies, of which 'immersion' is the most well-known, the anthropologist is able to engage in different styles of learning and understanding, acquiring discursive and practical knowledge, being simultaneously 'in' but not 'of' the other culture. As Hastrup and Hervik (1994:1–12) point out in an introduction to a recent collection of essays devoted to this issue, the process of acquiring that social knowledge in the field is central to the anthropological enterprise, yet we know surprisingly little about it: how understanding of the other occurs is insufficiently documented and still not fully understood.

There are, however, some who would deny that there is any mystery in the process, who would argue that there are, for example, procedures which allow you to conduct the research according to objective quasi-scientific procedures, hence the spate of publications on fieldwork techniques, statistical procedures, computer packages, a point nicely made by Karp and Kendall when they talk about the 'myth of fieldwork methods' (1982:251). Of course, no one would deny the benefits of judiciously using some of these advances in data-gathering techniques; the implicit criticism of the amateurism of an earlier generation of anthropologists who superciliously appeared to scorn method while producing, it is generally admitted, the classic and still (only just?) readable monographs, is readily accepted. Indeed many of us remember only too well how little prepared we were when we went into the field and how for a large part we were simply expected to pick things up as we went along.[2] The situation has changed in the last decade or so, especially in these days of quality control, but there is still a general feeling of dissatisfaction and uneasiness among students doing postgraduate courses in research methods and about to go into the field.[3] This uneasiness frequently gives rise to a lack of confidence both with respect to the quality and quantity of the material that students record in the field and subsequently in relation to writing up the material for a thesis or for publication. The reminder that we need to be more systematic in how we impart practical skills to students is, then, well taken, but techniques should not be regarded as an end in themselves, and we need to be equally forcefully reminded of what we hope to achieve by their use. There is a genuine danger that obsessive concern with technique can lead to the suppression of that point of departure so crucial to any successful anthropological enterprise: that other people experience the world in a way different from ourselves, and we can never be sure that we understand what the experience is which is masked or reflected by the behaviour which we are so busily monitoring and recording with our data collecting equipment. To give an over-simple example: how do we know what a smile means? Even in our own intimate circles can we always be sure that we know if it is expressing pleasure, contempt, bafflement, disdain, incomprehension, dismissal, anger, deviousness, distance, intimacy? And if we have such a problem within our own familiar, brightly lit[4] surroundings where we share a common set of habits, and when we have undergone years of training in how to decode social cues, how much more perplexing is the situation we face in the dimmed lighting of an alien culture?

There are two levels at which this problem of understanding is crucial in the field: one relates to the whole anthropological project of the moment; the other concerns the issue of simply coping with the quotidian reality of experience in the field. With regard to the first, there is now general agreement about the way in which fieldwork determines observation. It was again Evans-Pritchard (1976:241) who made the point that the best preparation for going into the field was a thorough grounding in the theory of social anthropology, since it was only through familiarity with such theory that one could properly direct one's observations and theoretically contextualise the impressions one was observing in the field. Regarding the second, there has perhaps not been so much discussion as the matter has warranted. What is at issue is the way in which we unavoidably bring into our understanding of what goes on around us those codes and conventions which we are familiar with from our own culture and to which what we observe seems most assimilable. In other words, we use ourselves and our own personal experience as primary research tools.

We are, however, rarely conscious that we are playing this dual role of investigator and instrument, that is, we rarely step outside ourselves in order to reflect on how our own life-histories are contributing to the perspectives we are accumulating: the reflexivity, if it comes, usually comes later when, self-consciously practising our profession, we write our ethnographies. In the field the professional and the personal are fused and we unconsciously engage in the process of making sense by assimilating. This process usually begins with dissonance, as time and time again in those first few months in the field we encounter unsettling situations in which we discover that our perceptions and expectations do not in fact measure up to the events which occur. The process ends in consonance, when we discover that our predictions are more likely to come out correctly because we have by this time learned new codes. Even at the end of our fieldwork, however, we may never be entirely confident that interpretations are unassailably correct.

This point about the difficulty of being sure of one's interpretation has been made many times before, most notably by Clifford Geertz in his account of ethnography as 'thick description', that detailed explanation of the symbolic actions of speech and gesture which allows us to understand the meaning which actors ascribe to their performance, enabling us, to take his famous example drawn from Ryle, to distinguish between a wink, a nervous tic of the eye and a parody of a wink (Geertz 1973b).[5] The point, however, needs to be stressed, not least because along with some well-taken criticisms of Geertz in relation to his practice, there is sometimes a failure to acknowledge the substantial contributions he has made to anthropology by his insistence on the discipline being largely one of interpretation. What we are doing in the field, then, to repeat, is making ourselves inward with a culture to the point where we feel as comfortable with it as we do with our own: where we have reached that degree of confidence that we can assure ourselves that even if we do not know the meaning of this one particular smile of the moment, we know at least the range it can encompass – we know how many types of ambiguity there are in the text.

The enterprise is of course impossible. To keep to the textual analogy for the moment, F.R. Leavis constantly used to deny the possibility of a literary critic dealing adequately with a literature written in a language other than his own, the implication being that the subtleties and nuances of the language, as well as a text's embeddedness within a historical tradition of language to which one begins to attune oneself from the instant of recognising the words, phrases, sentences and songs with which one is addressed from birth onwards, are impossible to grasp unless one is born to it. *Mutatis mutandis*, anthropologists going to the field relatively late in life must realise their limitations and the impossibility of total comprehension while still striving to accomplish the ideal. Inevitably there will be areas of knowledge which lie beyond their range, which will at times inhibit the grasp of the full meaning. Having sounded this note of caution with respect to the limits of ethnographic competence, we must, however, give due credit to the fact that a substantial amount can be shared, understood and passed on.

The point, though, is how? Are there techniques for facilitating the process of understanding of meaning as well as those for recording the behaviour which is an intentional expression of that meaning? Or is the process elusive, a question of individual temperament? One of the ways of explaining the procedure, indeed an explanation which is sometimes translated into specific advice, suggests that we regard the anthropologist as a role-player: child, stranger, friend, fictive-kinsperson are the usual roles mentioned, each requiring different behavioural attitudes and each allowing access to different realms of experience within the community. Another well-worn analogy – one which Andrew Beatty takes up in his contribution to the present book – is that of second-language learner. We acquire a language by gradual stages, becoming progressively more fluent until we feel we are fully in possession of it, with an implicit knowledge of the grammar, syntax, phonetics, idiom and vocabulary of the language, allowing us to reproduce it and understand it as though a native speaker. So we try to do the same as anthropologists with the culture and organisation of a society, learning how to act, recognising and responding to cues, positioning ourselves appropriately on the stage and being sensitive to the demands of ensemble performance. In Hastrup and Hervik's succinct formulation, 'Learning culture is a process of gradual familiarization in practice' (1994:7).

The analogy with language learning is attractive and we could develop it further, noting for example that there are certain people who are said to have a gift for languages, who relatively quickly acquire mastery of a language, whereas others employing the same learning techniques and often investing considerably more time and labour, never achieve the same skill: the pronunciation is not right, the syntax lapses, the idiom is stilted. Can we say the same of anthropologists in the field: that however conscientious the practice of techniques and the use of learning aids, some will inevitably, as a result of temperament, idiosyncratic quirks, historical accidents of personal experience, background, psychological make-up, be better than others at the task of understanding other cultures? And note this is a question of understanding, not simply mimicking, although mimicry is often a good indicator. A second-language speaker who comes close to native competence is not simply one who can imitate correctly: she also understands, since if she did not she would be unable to perform, that is, respond fully to cues which are not simply verbally marked but are semantically loaded with meaning. My feeling, then, is that the analogy here holds good. However conversant we may be with the various strategies for acquiring ethnographic understanding, however good at playing the various roles which may be foisted on us, or which we indeed choose deliberately to assume, some of us will be better learners than others.

I hasten to make two points at this stage. The first is that the example of language learning is an analogy and it is not intended to describe the process by which one acquires a knowledge of another culture, and that furthermore it is possible to be a very good ethnographer while being a poor linguist. The skills required for each task are different: an inability to reproduce with total accuracy Cantonese tones in speech is not a reflection of poor comprehension of the culture.[6] A second point is that the quick rapport and seeming inwardness which a good ethnographer establishes in the field does not guarantee the makings of a good anthropologist. Understanding is only part of the task, translating is the other. The most assimilated ethnographer in her writing often fails to match the arduous and painstaking work of the marginal anthropologist: indeed it is this awareness that underlies the warnings about 'going native'[7] or knowing too much.[8]

If, as all this suggests, data-collection techniques and pre-fieldwork advice on research methods must be understood as subject to the constraints and limitations of the personality of the individual anthropologist, what can one say about being in the field which will both acknowledge the bedrock mystery and serendipitous quality of the encounter with other people, and yet at the same time bring out something of the universal quality of the experience to which others – students, non-anthropologists – can vicariously respond? The answer lies to hand in this book: six personal accounts written not to advise or to demonstrate good practice, but precisely to illustrate the way in which knowledge and understanding grow out of prolonged encounters which elicit shifting perceptions of social reality, leading never to finality but to an ever-increasing awareness of the subtlety of interpersonal understanding.

'ANOTHER BOOK ON THE EXPERIENCE OF FIELDWORK?'

The origins of the book lie in a series of seminars I convened in the Anthropology Department of the University of Kent at Canterbury. At the time I had become increasingly persuaded that it was important to get away from the mechanical approach to anthropology which seemed to characterise our teaching. We were all aware that students, in particular postgraduate students, whether they intended to do fieldwork themselves or not, very much enjoyed personal accounts of fieldwork and we encouraged each other to leaven our lectures and seminars with 'tales from the field'. However, we all found that it really was not quite so easy as that. We might be happy to produce examples from our own observations to substantiate theoretical points – having done fieldwork among the Minangkabau, for example, I could rattle off at the drop of a hat beautiful examples of mother-in-law avoidance, uxorilocal residence and the nature of MB–ZS relationships – but when it came to talking about personal experience, emotions, relationships in the field, we were all rather tongue-tied. Our hesitancy had something to do with British teaching styles, particularly where anthropology is concerned, where the intention is to put across a substantial amount of theoretical knowledge organised around the key domains – kinship, politics, economics and religion – and drawing on classic examples. This approach inevitably shunts to the background any prolonged discussion of one's own material. It is also, I think, a product of genuine embarrassment and a reluctance to talk about oneself, partly out of modesty, partly out of fear for the vulnerability which comes from self-exposure. In part, the unwillingness to talk about oneself also stems from a strong commitment which still exists in social anthropology departments to the idea of the subject as a generalising science, from which it is essential to exclude those elements which would risk allowing it to drift in the direction of a humanities discipline.

Personally I have little sympathy with the last objection. Anthropology can encompass a wide range of disciplinary approaches including those from the natural sciences, but I myself veer to that kind of anthropology which derives its satisfactions from bringing about the recognition that 'all people in the four oceans are kin', and accomplishing that by the detailed focus on understanding the actions of individuals cross-culturally. On the other hand, I share that feeling of embarrassment at self-exposure – I don't, unless I'm in my cups, like to talk about myself – and I do feel that as teachers we have a responsibility to teach a corpus of knowledge to our students before we indulgently rehearse the minutiae of our own work. In general, then, I recognised the problem of describing the experience of being in the field, but I was sometimes frustrated at our failure to put across something of the excitement of fieldwork to our students. The possibility of organising a seminar series on the subject seemed to me to provide an initial opportunity at least to stir the imagination.

In thinking about what I wanted contributors to talk about within the gamut of possibilities which the topic 'fieldwork' suggests, I had in mind two very specific themes. When I first began to study anthropology systematically, I had mentioned in casual conversation at a party how dry and impersonal I found many of the monographs I had to read and how they lacked that crackle and sparkle which I knew from my own encounters with individuals very different from myself whom I had got to know well. In reply, in a vague almost off-hand sort of way which characterised his manner, my fellow-conversationalist asked whether I'd read Casagrande's (1960) edited volume *In the Company of Man*. When I said I hadn't heard of it, he suggested I dig it out of the library since it might be what I was looking for. I'm eternally grateful for the suggestion. The book, taken out of Haddon Library at Cambridge the next day, restored my flagging enthusiasm for anthropology and remains today a source of inspiration. The wonderfully evocative accounts in that book describe the relationship between anthropologists in the field and individuals whom they get to know well as either informants or friends or both, and in the telling of the relationship the reader is provided with extraordinarily graphic insights into the way in which two personalities, those of the anthropologist and the informant, in their mutual exchange of views and opinions grow together in maturity and both extend their understanding of what it is to be human. One can understand from these descriptions exactly how it is that Boissevain in a different context and unashamed by the potential charge of romanticism can be led to write '... it is the close and prolonged periods of contact with people in the field that stand out as high points in my life' (1970:71), or Read can say:

... gradually I began to respond to the villagers as individuals. It is not possible to say when this first occurred (perhaps when I found myself sitting with a man in the evening and realized that I had sought him simply for companionship and not because I wanted information), but the discovery has remained one of the most rewarding in the my life. I realize now that it is one of the benefits of my profession to experience this response to persons whose outlook and background could hardly be more dissimilar from my own. (1965:6)

This type of account detailing a special relationship with one informant from whom the anthropologist had gained her bearings within the society, was one alternative I mentioned to contributors in my invitation to them in which I referred specifically to Casagrande. None of them, however, took up this possibility, at least directly – Beatty's account does describe an informant but it is a negative portrayal – perhaps because Casagrande's book is not very well known here and it is not easy to obtain from libraries. My other suggestion bore more fruit. I referred to Geertz's essay 'Deep Play: Notes on the Balinese Cockfight' (1973c) and reminded contributors of the opening sequence – the cinematic term fits the style of the narrative – describing Geertz's entrée into Balinese society as he flees from witnessing an illegal cockfight and is given temporary asylum in the house of a stranger. It is a highly dramatic description and the intention underlying it is to focus sharply on that moment of acceptance into the society, after which – or so it would seem – the process of understanding which had been hampered up till then proceeded smoothly and apace. It was then an important 'moment' in the field, and I asked contributors to consider whether they could describe similar experiences of their own. As we shall see, this reference to Geertz did provoke reactions, not all of them positive.

Before referring specifically to the chapters, let me say something in reply to an objection voiced by a number of my colleagues to the whole enterprise of a series of chapters on personal experience in fieldwork. In the strong form this objection declares that this personal confession is at best simply embarrassing, is usually irrelevant navel-gazing and at worst is neurotic logorrhea: when our interest is substantive anthropology, why should we want to know about the personality of the anthropologist? In the weak form there is some acknowledgement that a description of how ethnographic data is acquired and anthropological knowledge generated is useful, but we have now reached the stage where there is a glut of such accounts and it is time to call a halt and get back to our proper job.[9]

Of course my reply to these objections was robust and it sprang from a very diverging view of both what the potential scope of anthropology is and the relative importance of accounts which site the anthropological persona squarely in the text. There is no need to go over the general arguments again: they are now very familiar in the anthropological literature – for a good overview of the discussion see Marcus and Cushman (1982), Clifford and Marcus (1986) and for more recent arguments Nencel and Pels (1991) and Cohen (1992) – and a general consensus has emerged in the groves of the anthropological academy allowing people to live in peaceful coexistence in their departments. I would, however, like to offer some justification to those who wearily throw up their hands at 'yet another book of subjective impressions of fieldwork'.

My initial response is to throw out the counter-challenge and ask 'Well where are these books?' And when the usual list is provided, I reply that I am happy to accept that the concern for describing the process of fieldwork is not a recent phenomenon, that even before the new ethnography we had books such as Hortense Powdermaker's *Stranger and Friend* (1967) and Elenore Bowen's *Return to Laughter* (1956), but note how frequently these books come up as examples and note how little else there is from this earlier generation. More recently, there have been excellent collections such as the volume edited by George Spindler (1970) entitled *Being an Anthropologist* which is a set of personal accounts linked to monographs published in the well-known Holt, Rinehart and Winston series. Written in the last 15 years, it is true, we have seen some wonderfully reflexive accounts of which perhaps the most well-known – notorious in the eyes of some of my colleagues – are Paul Rabinow's *Reflections on Fieldwork in Morocco* (1977) and Jean-Paul Dumont's *The Headman and I* (1978). Even more recently we now have a book about the issue of sexual relationships in the field edited by Kulick and Willson (1995) the very title of which, *Taboo*, indicates the detached private and set-apart nature of the subject which had never previously been considered suitable for public disclosure.

Let me make two observations at this point. The first is that none of these books is British.[10] All of them take as their point of orientation American anthropology. In terms of the way the narratives are constructed and the issues are framed, the single books or contributors to collections are implicitly written to respond to American experience, and the implicit comparison is with American society inside and outside the university. What we have from British anthropologists is relatively meagre. The point of referring to a specifically British style of tradition of anthropology is not to endorse national stereotypes, or some caricature of what constitutes British or American personality, but to draw attention to broad ideas of what is culturally acceptable in terms of self-disclosure. It has been argued in this respect that British autobiographical styles in general in the past decade have become more openly confessional than they were in the past, with men for example being far more explicit about their emotions than is customary. Blake Morrison's *And when did you last see your father?* (1993) is often taken as one of the best examples of this new writing. Now irrespective of whether this writing does in fact represent a change in cultural attitudes, the very fact that it is perceived to do so, suggests a drawing of boundaries around culturally acceptable notions of self-disclosure. The British confessional style is more ironic, more detached, never in fact being explicit, and frequently obscuring emotion entirely behind self-mockery and humour. This explains, it seems to me, why the nearest that we have come so far to self-reflexive accounts in Britain has been Nigel Barley's book (1986) which, however great its popular appeal, is received by most British anthropologists with an angry sense of 'no, that is not it, at all'. The only recent book which does seriously attempt to address the issues in British anthropology is *Autobiography and Anthropology* edited by Okely and Callaway (1992), all the contributions to which – with one possible exception – put beautifully into context, temporal and academic, the conditions under which British-trained anthropologists have recently engaged in fieldwork.

And it is the word 'recently' which I wish to take up in making my second observation. In comparing the several accounts of fieldwork, a noticeable feature that stands out is the striking reorientation of the focus of the works from period to period. Three particular differences immediately come to mind: the difference in the selection of issues considered worthy of attention, the positioning of the anthropologist within the text and the relevance of the immediate geo-political context in determining the composition of the narrative. Without going into any detail let me simply illustrate briefly what I have in mind. Regarding the choice of issues, one might note that for the earlier generation of anthropologists, physical deprivation and intellectual isolation was clearly a major concern which preyed on anthropologists' sense of identity both at the time – think of Malinowski's diaries – and on subsequent reflection – think of Bowen's descriptions of food. For later anthropologists, particularly for the most recent generation, these issues are not a problem: this is largely because global circumstances have changed and there is consequently no real feeling of isolation, and partly, I think, it is because a younger generation is probably better physically and psychologically prepared to put up with discomfort. It is other features of their day-to-day lives in the field which exercise them more. One obvious one is gender relations, where what is encountered in the field differs so starkly from what at least is preached in their own milieux.

With regard to the positioning of the anthropologist, what I have in mind is the degree to which the distance between the anthropologist and the members of the other culture becomes problematic in a way which does not seem to have been an issue for previous generations. This shouldn't be taken to mean that earlier generations were somehow more aloof or remote, less open to accepting intimacy with others: the accounts of older anthropologists in Casagrande's volume clearly refute this. It is rather a question of a deliberately maintained distance, an awareness of unbridgeable difference recognised on both sides between anthropologist and informants. Whether one has in mind Evans-Pritchard or Elenore Bowen or even from a more recent generation Gerald Berreman (1962) or Margery Wolf (1968), it is clear that anthropologists feel secure and comfortable in their role of professional stranger which, even if it requires constant explanation, does not demand self-justification: in other words the anthropologist's role is taken for granted by the anthropologist. That, it seems to me, is much less the case for the most recent generation of anthropologists where the question frequently arises: what am I doing here? There are various ways one might account for this new self-questioning: it might well be the product of the success of anthropology in terms of the dissemination of knowledge of the other to the point that an anthropologist feels much closer in sympathy and spirit to those among whom she works. The consequence of this feeling is that the probing research she is required to undertake seems now much more invasive and intrusive: the embarrassment of asking personal questions is directly proportional to the degree of familiarity with which one knows one's informant. At almost the other extreme of self-consciousness there is also, I think, a concern that, however the Western anthropologist tries to purge herself of the last vestiges of nineteenth-century romanticism underlying the desire to do fieldwork in exotic places, there still lingers a suspect intellectual voyeurism. Another explanation, however, might account for the uneasiness by reference to changing global circumstances, the third of those points of difference noted above.

The aftermath of independence and the restructuring of the global economy have of course had consequences which have affected both students in the academy and hunter-gatherers, even in the remotest parts of the world. During the colonial period, the nature of relationships between informants and anthropologists was, whether the anthropologist liked it or not or whether she was aware of it or not, determined by the structures of power and domination underlying international relations at the time. And one could argue that the hangover of a colonial mentality continued to affect relationships right up to the mid-1960s. After that time,[11] however, it was largely gone, and in its place arose a spirit of independence and equality, which, although it might have been predicated on a false belief that the new world order did confer equality, did affect the way in which all at the levels from government bureaucracies right down to the lowest village levels anthropologists were perceived. In terms of our own preoccupations here with the changing context in which anthropology is conducted, the consequences of this change are reflected at a trivial level in those anthropologists' tales which deal with the nightmares of obtaining bureaucratic permissions to do the research or reside in the country, or, at the other extreme, desperately trying to avoid being taken for hippies: two types of concern which are simply absent from these earlier accounts of fieldwork.

Each of these dimensions of the changing self-perceptions of the anthropologist requires much more detailed treatment and analysis. (For a much fuller discussion of some of the issues see the book edited by Fox (1991).) The reason for sketching them out here was to demonstrate briefly but, I hope, forcefully, that the objection to yet more accounts of fieldwork has failed to recognise that each account of fieldwork is valuable not only for the measure of personal reminiscence and reflection it contains, which may not be to everyone's taste, but also for the information it contains about the changing circumstantial context in which the discipline continues to operate which in turn affects the nature of the discipline itself. A critical analysis of the texts and careful comparison among them have a great deal to tell us about our discipline. (For starters, for example, someone might like to compare Bowen (1956) and Cesara (1982).) To that extent, then, we can never have enough accounts of fieldwork if we are to chart the history and development of the discipline, a point I am happy to see endorsed by Gluckman, in a relatively early – and, one might say for that generation, uncharacteristic – plea for more accounts of fieldwork (1967:xviii) and by Nencel and Pels who write that 'we need a lot more accounts from the grassroots to see where we stand'

(1991:17). For the early years of social anthropology, with the one or two exceptions already mentioned, we have to rely on archival material, letters, committee reports, government files, minutes of meetings and the oral histories derived from the senior generation. In years to come, in addition to this material, histories will also be able to turn to much more in the way of published autobiographies, fictionalised narratives and the kind of essays which are presented here.

DEGREES OF STRANGENESS

The chapters which follow are all personal accounts of fieldwork, but rather than deal comprehensively with the whole experience, each takes up particular episodes or scenes which were representative or have retained a powerful hold on the memory. Several contributors resisted the idea that fieldwork comprises epiphanic moments, when sudden flashes of insight illustrate what had previously been obscure. Gardner and Beatty explicitly reject this idea, perhaps sharing the scepticism of Richard Fox when he calls into question precisely how radical postmodernist critiques are: 'The magic of fieldwork, the epiphany of the fieldworker, the sink-or-swim during the anthropological novitiate – all these antique conceits of anthropology are now embellished and modernized by an ostensibly self-critical reflexivity' (1991:7). Indeed Gardner is dubious about the usefulness of compartmentalising fieldwork in a special category and argues strongly that to elaborate the process of understanding that occurs during fieldwork one must recount the pre-fieldwork and post-fieldwork contexts, and she demonstrates this in her own case by describing the way in which her feminist perspectives initially controlled her perceptions of women's lives in Sylhet.

Equally important are the post-fieldwork reflections on what precisely occurred during fieldwork. In a beautifully evocative chapter, Paul Spencer (1992) has shown how the memory plays tricks and colludes with the way in which one goes about the process of self-fashioning or constructing a text. There is the same nagging concern underlying Andrew Beatty's chapter when he writes about his experiences on the island of Nias off the west coast of Sumatra. A celebratory feast which the anthropologist holds to ease his passage into the community goes disastrously wrong, and years later aspects of it remain a puzzle.

For Allison James, writing of her experience doing research among primary school children in England, it is also the subsequent reflection on events which illuminates what occurred. Listening to a tape she understands what it was a small boy was trying desperately to convey to her about himself, and recognises, in a way which must give us all pause to shudder at how we ourselves have behaved, how insensitive she was to his pleas at the time.

For Anna Grimshaw, too, it is the vividly remembered scenes of her fieldwork in a Buddhist nunnery in the Himalayas as they recur to her long after which gives rise to her introspection. The chapter by Spencer referred to above concludes by stating that with the end of fieldwork came a decision to leave anthropology: the catharsis which he now recognises had been what prompted him into anthropology had been accomplished and he sets out to do something new. Grimshaw too felt that she had taken leave of anthropology, and her contribution is an attempt to explain her dissatisfaction, while at the same time bringing home to the reader how emotionally engaged she was in her fieldwork. My own chapter, the only one, incidentally, not actually given as a seminar – it is not something which I could have presented orally – also touches on this feeling of frustration that anthropology would not allow me to deal adequately with the personal and the private. My account of the death of a young girl from a village in Kerinci in the highlands of Sumatra represents an attempt to deal with that frustration while still being true to what gave rise to it: a desire to share an experience which, through the very fact of its being constructed to fit this form, risks shattering.

At the heart of my chapter and indeed of all the other chapters too, there is the evocation of friendships in the field, not single individuals of the kind who were the subjects of Casagrande's contributors, but none the less informants who became and have stayed friends. In Cris Shore's chapter we see how problematic the establishment of friendships can be. While trying to be all things to all men while investigating local politics in a town in Italy, Shore finds himself caught in an impossible situation. After one vividly remembered incident, the die is cast and Shore is committed to a group of friends from whom over the course of their friendship he learns as much about himself and his own political opinions as he does about small-town Italian politics.

Each of the chapters reflects not only the personal slant of each of the individual contributors in relation to how they have recalled the experience of fieldwork, they also reflect much more directly the individual personalities, predispositions and temperaments of the individuals themselves. This aspect of their accounts is not for me to comment on in any way and, besides, readers should be left to themselves to enjoy the encounter with the anthropologists. One or two general remarks, however, may be appropriate. All the contributors have been trained in departments of social anthropology in British universities, and the fieldwork experiences which are described by them all took place more or less within the twelve years between 1978 and 1990. A point to note here, particularly for those unfamiliar with the institutionalisation of social anthropology in British universities, is the relative homogeneity of the tradition of fieldwork and preparation for the field across university departments in this period. Because of its small size and very recent development as an academic discipline, social anthropology in Britain derives its practice and its approach from a common ancestry of only two generations' depth, and consequently differences in fieldwork method are not marked. After preparation lasting from six to nine months, researchers go out to the field as individuals working in isolation ideally by 'immersing' themselves in a community where they gather data from one to two years, after which they return to write up the monographs in a space of one to four years. On the other hand, the variety of the areas in which the fieldwork was done and the particular sub-disciplines within the subject which are represented here reflect the heterogeneity of the discipline. Anthropology in the 1980s in Britain was turning its attention much more to Britain and to Europe. Hence Allison James could without hesitation embark on anthropological research in an English school. On less familiar ground but still within Europe, Cris Shore first learns Italian and then embarks on his research. The highlands of Sumatra may seem very distant from Cambridge, but as I explain in my contribution the distance was in fact not such a great one. The other three contributors travelled a much further emotional, psychological and intellectual distance. Katy Gardner was able to use some of the linkages which connect Sylhet to Britain, but the world in which she found herself was none the less still quite alien to her. Andrew Beatty on Nias was also entering an alien environment, although one which is much less on the periphery than it was 50 years ago. Anna Grimshaw's experiences were apparently located at perhaps the greatest remove from the familiar. Not only was she secluded within a nunnery, but for a long time she was unknown to the government authorities.

Nevertheless, whatever the degree of initial strangeness in their encounters, all the contributors seem to achieve an intimacy with their surroundings and a closeness to people which, as I have suggested above, marks off the experience of post-1960s anthropologists very distinctively from their predecessors. As individuals they differ in the extent to which they make explicit how their formative educational and autobiographical experiences affected both their views of anthropology as a discipline – and therefore of what they hoped to achieve – and their capacity for relating to and engaging with their informants. Reading their accounts, however, makes very clear that none of them, however temperamentally they might have been attracted to anthropology, were ever under those romantic illusions of uncovering the exotic which exercised their predecessors. Power and inequality in relationships are issues of which they are acutely self-conscious. The lessons of the 1970s have been well learned, in so far as the 'other' in the narratives is transformed into one of 'us' at the same time as 'we' – vicariously through the agency of the anthropologist – are transformed into the 'other'. What makes this possible is not the forced approximation of us all to the same commonly shared symbols of superficial meaning, but a constantly rehearsed demonstration that interpretation does take place according to universal rules and translation does allow us to understand each other. For as long as this process of translation is deemed to be a pursuit worthy of our highest aspirations, I am in no doubt that anthropology will continue to be the leading humanist discipline.

NOTES

1. This notion is firmly stated by Epstein in his preface to *The Craft of Social Anthropology*, a classic manual on fieldwork techniques still much in use:

   The tradition of conducting fieldwork, usually in more or less isolated and 'exotic' communities, and the theoretical perspectives that stem from it, would probably count for many people as one of the major contributions of social anthropology to social science. Against this tradition, it is not surprising that preparation for fieldwork has come to be seen as an essential part of the training of students in the subject, and fieldwork itself as a unique and necessary experience, amounting to a kind of rite de passage by which the novice is transformed into the rounded anthropologist and initiated into the ranks of the profession. (1967:vii)

There are, however, those who would argue that too much is made of fieldwork as the hallmark of anthropology. Ingold, for example, in an attack on introducing fieldwork exercises into anthropology curricula reminds us that we spend most of our time as anthropologists in reading (1989:3). Needham seems to have similar misgivings when he writes – though not I should say in the context of remarks about fieldwork – 'Actually, we have altogether too many facts, so many that we do not know what to do with them; and whenever we fail in explanation or understanding it is not usually for a lack of sufficiency of ethnographic data' (1985:40). The implication seems to be that we would do well to spend more time on the intellectual analysis of what we have and be less obsessed by the collection of new data.

2. One sometimes gets the impression in relation to preparation for fieldwork that anthropologists as a collectivity are like those who are ignorant of history, ever bound to repeat the failings of their predecessors. After all Evans-Pritchard was complaining about the inadequacy of his preparation as early as the twenties:

... when I was a serious young student in London I thought I would try to get a few tips from experienced fieldworkers before setting out for Central Africa. I first sought advice from Westermarck. All I got from him was 'don't converse with an informant for more than twenty minutes because if you aren't bored by that time he will be'. Very good advice, even if somewhat inadequate. I sought instructions from Haddon, a man foremost in field-research. He told me that it was really all quite simple; one should always behave as a gentleman. Also very good advice. My teacher, Seligman, told me to take ten grains of quinine every night and to keep off women. The famous Egyptologist, Sir Flinders Petrie, just told me not to bother about drinking dirty water as one soon became immune to it. Finally, I asked Malinowski and was told not to be a bloody fool. (Evans-Pritchard 1976:240)

3. The seriousness of the concern became apparent in the Conference held for postgraduate anthropology students in Oxford in March 1998, the 'Marett Project', funded by the National Network for Teaching and Learning Anthropology. All participants expressed a deep disquiet about courses in research methods (personal com-

munication from two of the participants, and from the official observer Dr Stella Mascarenhas-Keyes).

4. To pursue a textual analogy here, I note that Hartman makes a similar point about uncertainty in literary analysis. He writes (quoting Stephen Booth's 'An Essay on Shakespeare's Sonnets', pp.186–7):

    Even when the reader has 'the comfort and security of a frame of reference ... the frames of reference are not constant, and their number seems limitless' ... What is inexplicit is as functional as what is patently there, though it is hard to describe in a rigorous manner the relation between what is provided and what elided, what is verbal and what situational, what is foregrounded and what understood. (Hartman 1980:266)

5. Hartman draws our attention to exactly the same phenomenon in noting that, 'A statement made in a novel, as compared to a statement made in a restaurant, is subject to a different kind of interpretation, but it is not a different statement. This look-alike (sound-alike) quality is disconcerting ...' (1980:265).

6. Margaret Mead had already in 1939 reacted against Malinowski's emphasis on fluency in the vernacular (see Clifford 1988:30) and had advanced a number of arguments in relation to working through trained interpreters (1964:34). However, the consensus of anthropological opinion appeared to be against her, particularly in British social anthropology. Recently, however, there appears to have been a shift to a more realistic perspective which understands the difficulties of acquiring native competence. Marcus and Cushman, for example, write: 'It may be that a total linguistic control is not necessary for ethnographic authority, but rather just control of that part of the language that informs a defined task of interpretation' (1982:36). For my own part I am of the Malinowskian persuasion and believe that one should aim for native competence while at the same time acknowledging that one frequently falls short of the ideal.

7. With respect to 'going native', the spectre of the ethnologist Frank Cushing who lived for years among the Zuni of North America continues to haunt American anthropology. As Clifford puts it: 'Cushing's intuitive excessively personal understandings of the Zuni could not confer scientific authority' (1988:28). Is it the memory of Cushing's ghost, perhaps – along with a certain scepticism –

which makes us so uneasy about our own contemporaries – Carlos Castaneda, Florinda Donner – who become so absorbed into the culture of others?

8. On the dangers of knowing too much, there is a lot of anecdotal evidence heard in postgraduate seminar rooms about X or Y who had lived too long in a certain society and knew it so well that he suffered from intellectual paralysis every time he sat down to write about it. Gluckman makes the same point: '[The anthropologist] has continually to clarify his own role in the society he is studying so that he is neither left completely in ignorance on the outside nor completely captured at the centre, where he knows so much that he can publish nothing' (1967:xviii).

9. With reference to an earlier period in the recent history of anthropology, Needham makes this same point. He writes about wanting to make

    ... a protest against the vogue of unapplied methodology, programs for hypothetically superior research, evaluating surveys of work done by others, introductions to social anthropology, conferences, and all such instruments and occasions for methodological pronouncements. It [the essay he has written] is a plea for getting on with the job ... (1962:vii)

    None the less, as the author of *Exemplars*, I hope he might be in some sympathy with the present collection of essays.

10. There are, however, two excellent collections of essays on the experience of fieldwork published in India, Béteille and Madan (1975) and Srinivas et al. (1979), which capture exactly that kind of description of the intellectual and personal contexts in which fieldwork was conducted. The essay by Madan (1975) is particularly valuable for describing the influence of British social anthropological traditions on his thinking.

11. Perhaps the change had already occurred a little earlier. I was struck recently when rereading Gluckman (1967) to note how his remarks on the global changes of the late 1950s having affected relations between anthropologists and informants seemed to anticipate many later critiques: 'The tribal people who have provided anthropologists with so much other data are now often able and eager to read what is being said about themselves. They are likely to protest if they think that they are being misrepresented ...' (1967:xviii).

REFERENCES

Barley, N. (1986) *The Innocent Anthropologist: notes from a mud hut*, Harmondsworth: Penguin Books (first edn 1983).

Berreman, Gerald D. (1962) *Behind Many Masks: Ethnography and Impression Management in a Himalayan Village*, Monograph Number 4, The Society for Applied Anthropology. Rand Hall, Cornell University, Ithaca, NY: The Society for Applied Technology.

Béteille, André and Madan, T.N. (eds) (1975) *Encounter and Experience. Personal Accounts of Fieldwork*, Delhi: Vikas.

Boissevain, Jeremy F. (1970) 'Fieldwork in Malta' in Spindler (ed.) (1970).

Bowen, Elenore Smith (1956) *Return to Laughter*, London: Readers Union and Victor Gollancz.

Caplan, Pat (1992) 'Spirits and sex: a Swahili informant and his diary', in Okely and Callaway (1992).

Casagrande, Joseph B. (1960) *In the Company of Man. Twenty Portraits by Anthropologists*, New York: Harper and Brothers Publishers.

Cesara, Manda (1982) *Reflections of a Woman Anthropologist*, London: Academic Press.

Clifford, James (1988) *The Predicament of Culture*, Cambridge, MA and London: Harvard University Press.

Clifford, James and Marcus, George E. (eds) (1986) *Writing Culture*, Berkeley, CA and London: University of California Press.

Cohen, Anthony P. (1992) 'Self-Conscious Anthropology', in Okely and Callaway (1992).

Cohen, Anthony P. (1994) *Self Consciousness*, London: Routledge.

Diamond, Stanley (1974) *In Search of the Primitive. A Critique of Civilization*, New Brunswick, NJ: Transaction Books.

Dumont, Jean-Paul (1978) *The Headman and I*, Austin, TX: University of Texas Press.

Epstein, A.L. (ed.) (1967) *The Craft of Social Anthropology*, London: Tavistock.

Evans-Pritchard, E.E. (1976) *Witchcraft, Oracles and Magic among the Azande* (abridged edn with an introduction by Eva Gillies), Appendix IV 'Some Reminiscences and Reflections on Fieldwork', pp. 240–54, Oxford: Clarendon Press.

Fox, Richard G. (ed.) (1991) *Recapturing Anthropology. Working in the Present*, Santa Fe, NM: School of American Research Press.

Geertz, Clifford (1973a) *The Interpretation of Cultures*, New York: Basic Books.

Geertz, Clifford (1973b) 'Thick Description: Toward an Interpretive Theory of Culture', in Geertz (1973a).

Geertz, Clifford (1973c) 'Deep Play: Notes on the Balinese Cockfight', in Geertz (1973a).

Geertz, Clifford (1988) *Works and Lives. The Anthropologist as Author*, Oxford: Polity Press.

Gluckman, M. (1967) Introduction in Epstein (ed.) 1967.

Gould, Harold A. (1975) 'Two Decades of Fieldwork in India – Some Reflections', in Béteille and Madan (1975).

Hartman, Geoffrey, H. (1980) *Criticism in the Wilderness*, New Haven, CT: Yale University Press.

Hastrup, Kirsten and Hervik, Peter (eds) (1994) *Social Experience and Anthropological Knowledge*, London: Routledge.

Ingold, T. (1989) 'Fieldwork in Undergraduate Anthropology: An Opposing View', *British Association for Social Anthropology in Policy and Practice Newsletter*, 3 (Summer): 3–4.

Karp, Ivan and Kendall, Martha B. (1982) 'Reflexivity in Field Work', in Secord (1982).

Kulick, Don and Willson, Margaret (eds) (1995) *Taboo: sex, identity and erotic subjectivity in anthropological fieldwork*, London: Routledge.

Madan, T.N. (1975) 'On Living Intimately with Strangers', in Béteille and Madan (1975).

Marcus, George, E. and Cushman, Dick (1982) 'Ethnographies as Texts', in Siegel, B. (ed.) *Annual Review of Anthropology 1982*, 11: 25–69.

Mead, Margaret (1964) 'Native Languages as Field-Work Tools', in *Anthropology. A Human Science. Selected Papers, 1939–1960*, Princeton, NJ: D. van Nostrand, 15–36 originally published in *American Anthropologist*, 41 (2), 1939.

Morrison, Blake (1993) *And when did you last see your father?*, London: Granta Books in association with Penguin.

Needham, Rodney (1962) *Structure and Sentiment. A Test Case in Social Anthropology*, Chicago and London: University of Chicago Press.

Needham, Rodney (1985) *Exemplars*, Berkeley, CA and London: University of California Press.

Nencel, Lorraine and Pels, Peter (eds) (1991) *Constructing Knowledge, Authority and Critique in Social Science*, London: Sage.

Okely, Judith and Callaway, Helen (eds) (1992) *Anthropology and Autobiography*, London: Routledge.

Powdermaker, Hortense (1967) *Stranger and Friend: The Way of an Anthropologist*, London: Secker and Warburg.

Rabinow, Paul (1977) *Reflections on Fieldwork in Morocco*, Berkeley, CA and London: University of California Press.

Read, Kenneth E. (1965) *The High Valley*, New York: Charles Scribner's Sons.

Richards, David (1994) *Masks of Difference. Cultural representations in literature, anthropology and art*, Cambridge: Cambridge University Press.

Rynkiewich, Michael A. and Spradley, James P. (eds) (1981) *Ethics and Anthropology. Dilemmas in Fieldwork*, reprint edn, Malabar, FL: Robert E. Krieger.

Secord, Paul F. (ed.) (1982) *Explaining Human Behaviour. Consciousness, Human Action and Social Structure*, Beverly Hills: Sage Publications.

Spencer, Paul (1992) 'Automythologies and the reconstruction of ageing', in Okely and Callaway (1992).

Spindler, George D. (ed.) (1970) *Being an Anthropologist Fieldwork in Eleven Cultures*, New York: Holt, Rinehart and Winston.

Srinivas, M.N., Shah, A.M. and Ramaswamy, E.A. (eds) (1979) *The Fieldworker and the field: problems and challenges in sociological investigation*, New Delhi: Oxford University Press.

Wolf, Margery (1968) *The House of Lim*, New York: Appleton – Century – Crofts.

1 FICTIONS OF FIELDWORK: DEPICTING THE 'SELF' IN ETHNOGRAPHIC WRITING (ITALY)

Cris Shore

RETHINKING THE CONCEPT OF 'FIELDWORK'

The aim of this chapter is to reflect analytically on the fieldwork encounter, particularly those frequently cited 'critical experiences' during fieldwork (or so they sometimes appear with hindsight), when a situation suddenly becomes clear, or when anthropologists are forced to radically reflect on and reconsider their research project. What follows is an attempt to describe and analyse some of the formative events and influences that shaped my own fieldwork study of the Italian Communist Party (PCI) in the early 1980s. I want to illustrate the ways in which – and the mechanisms through which – those experiences generated critical anthropological insights, both personal and professional. First, however, I wish to contextualise this by raising some questions concerning the way fieldwork has traditionally been conceptualised in anthropology, and the problems this poses for anthropological theory and practice. Thinking about my own 'critical experiences' led me to question not only certain assumptions about the nature of this activity that we call 'fieldwork', but also the ways in which anthropologists represent (and often misrepresent) themselves within their ethnographic accounts. I also began to ponder why British anthropologists have traditionally been so reluctant to write candidly about the more personal aspects of their fieldwork encounters. In American anthropology, by contrast, the 'self-reflexive', autobiographical account of fieldwork experiences has emerged as a distinct ethnographic genre (Stocking 1992:13; cf. Geertz 1973; Rabinow 1977). Rereading these texts, however, I found myself questioning whether writing ourselves into our ethnographic scripts really does lead to a more complete, accurate or honest ethnographic picture, and why this self-reflexive act of 'authoring ourselves' is more problematic than it may seem (hence the title of this chapter).

Evaluating my own fieldwork leads me to conclude that there have been many of those 'critical experiences', but both before and after, as well as during, fieldwork. In fact, it is no longer clear to me where 'fieldwork' proper begins or ends. The idea of 'the field' as a discreet, bounded geographical locale is proving to be increasingly outdated and untenable (Kohn 1995; Fox 1991; Wolf 1982).[1] Anthropology as a discipline has only barely begun to rethink the way we conceptualise fieldwork. The traditional idea that fieldwork can be neatly divorced or 'bracketed off' from 'normal' time and space is itself a highly dubious and problematic notion. This model, reminiscent of the anthropological concept of the 'liminal phase' in the rites of passage, may be useful for the purposes of constructing research grants and timetables, but it is not how fieldwork proper is experienced in practice. In my case, like many others, doing participant research in 'the field' was only one part of the fieldwork process. Although I did not end up marrying any of my fieldwork informants, as anthropologists often do, the close personal ties and cultural exchanges created in 'the field' have continued to this day.

Despite the lingering image of anthropology as an artisan craft based upon exotic and intense fieldwork encounters overseas, most anthropological work actually takes place in rather humdrum, non-exotic places outside or away from the site of ethnographic fieldwork (Fox 1991), for example, in libraries and lecture halls, at conferences, at the departmental seminar, on the word processor. It also takes place increasingly under 'factory-type' conditions of university discipline (that is, negotiated research leave or unpaid sabbaticals, pressure to score a higher mark in the next Research Assessment Exercise, proposals dashed out to meet Research Council closing dates). University life is also part of the fieldwork context. Moreover, many of the major ethnographic insights dawn on one gradually and slowly – often in the writing-up phase, or what Cohen (1993) calls the 'Post-fieldwork fieldwork' period. Fieldwork is a process as well as a series of chaotic encounters, and while there may be many revelatory fieldwork experiences, those revelations often come to us slowly and methodically. To put it in metaphorical terms, it is not the 'Road to Damascus' but the 'Road to Wigan Pier': not the effect of a sudden, blinding flash of light, but the much slower destination reached through sore feet and blisters.

PROBLEMS WITH THE NOTION OF 'REFLEXIVITY'

Let me begin, therefore, by raising a few questions about 'reflexivity' and the way anthropologists situate themselves in relation to their fieldwork. It is hard to talk about the significance of fieldwork in any meaningful sense without being highly personal and a little indiscreet. Being invited to write this chapter made me conscious of the fact that I have never discussed the nitty-gritty details of my fieldwork experience before, or at least not in public or to an audience of fellow anthropologists. I am not alone in this. Many anthropologists are quite guarded about 'opening themselves up' to the critical gaze of their colleagues by discussing the precise details of how they carried out their research (or were unable to, as the case may be). Most socio-cultural anthropologists, says Armstrong (1994:xix), 'modelling the writing of those by whom we have been trained, hide who we are and the roles we play in our research when writing for our professional colleagues'. The more honest, personal and 'reflexive' accounts tend to appear later, but these are often written under a pseudonym, or are classified as non-anthropological texts.

The reasons for this discretion are not difficult to fathom. It stems partly from a desire to maintain one's professional integrity and authority (many fieldwork experiences are confidential and could be ethically compromising[2]), and partly from the view that the object of anthropological interest should not be ourselves but the peoples we study. Reluctance to reveal details about the methods we use may also stem from the uncertainty and insecurity many anthropologists feel about those methods. As most anthropologists know, the term 'intensive fieldwork' (or 'qualitative research') is a gloss that covers a vast array of promiscuous techniques and messy encounters, with 'data' often culled from the most unlikely and improbable sources. Very often too, the anthropologist has received little or no training in ethnographic fieldwork techniques. Indeed, this is one of the central paradoxes of anthropology: as Stocking (1992:13) points out, while ethnographic fieldwork is 'virtually a sine qua non for full status as an anthropologist' – serving as the discipline's 'central tribal rite' of initiation – the same cannot be said of formal fieldwork training (more on this below). The 'conspiracy of silence' about fieldwork, as it sometimes appears to postgraduate students and 'outsiders', would appear to function as a screen for protecting both the discipline's professional image *vis-à-vis* outsiders, and its mystique as far as initiates are concerned.

Reluctance to write about the details of one's personal fieldwork encounter can also be attributed to the history of the discipline in Britain, its structural-functionalist legacy and, more specifically, the limited definition of what constitutes 'the science of anthropology' bequeathed by Radcliffe-Brown and Malinowski (whose shocking fieldwork diaries were, significantly, only published posthumously in 1967). It is interesting to compare British and American anthropology in this respect. The relative acceptance and growth of reflexive anthropology in the US comes from a long-standing connection between cultural anthropology and the humanities/literary studies and a far greater degree of sympathy for psychology (for example, Whorf, Mead, Spiro, Hallowell). When literary studies in the 1980s turned towards texts and the self, anthropology followed. British anthropology, by contrast, held to a rigidly Durkheimian self-image, one that was positivistic but strongly anti-psychological, that defined anthropology as the scientific study of 'other' cultures. British anthropologists were therefore strongly discouraged from introducing their own 'subjectivities' into the frame of analysis, or from reflecting too critically on their own theoretical orientation. And yet we know that the learning process is based precisely on subjective, personal encounters. Moreover, fieldwork is an emotional encounter as well as an intellectual exercise – something which has not always been properly recognised in British anthropology.

Nowadays, however, an element of reflexivity has become almost *de rigueur* in most ethnographic writing. This is to be welcomed: some celebrate it as marking a paradigm shift or definitive break with 'sterile scientism' (Grimshaw and Hart 1995). Reflexivity, however, is more multifaceted and problematic than it appears. The paradox of anthropology today is that at the same time that feminism and the so-called 'literary turn' (anthropology's answer to postmodernism) have given self-reflexive writing a measure of respectability and legitimacy, anthropologists have become increasingly self-conscious and uncomfortable about writing themselves into their texts. Part of the reason for this is the postmodern critique which has shown us that writing ethnography is an assertion of power and a claim to authority. We now see clearly the artifice and literary conventions behind Evans-Pritchard's ethnography – what Geertz (1988) calls his authorial 'signature': how he first personified himself among the Nuer only to then 'disappear' in favour of a detached, scientific omniscience for the remainder of the book. As Richard Fox (1991:6) points out, we are now much more suspicious of ethnography's claim to provide tidy portraits of the Other, and 'wary of the conventions by which we convince readers that we were really there and faithfully got the native point of view'. Fox's point is that these accounts 'represent a hackneyed genre even when they happened'. So should anthropologists respond to the postmodern critique by writing more self-reflexive, subjective and experimental ethnographies (perhaps of the kind suggested by Stephen Tyler (1986)), and if so, what are the problems of trying to write ourselves into our own narratives?

One of the common charges levelled against anthropology's 'reflexivists' (or postmodernists) is 'solipsism' or self-indulgence – even an 'egocentric and nihilistic celebration of the ethnographer as author, creator and consumer of the Other' (Polier and Roseberry 1989: 236), and a narcissism that results in 'inadvertent ethnocide' (Lewis 1985:369). Solipsism, to give it its dictionary definition, is 'the philosophical view that the self is all that exists or can be known'. Anthropologists may not agree with the first part of that sentence but there is a serious theoretical debate about whether we can ever really 'know' other people's subjectivities or whether we can ever escape from the prison-house of our own linguistic conventions, concepts and protocols. Some writers therefore suggest that anthropology should abandon its pretence of being a social science and its quest to 'represent the other' since this has now been 'exposed' as part of a failed neo-colonialist, orientalist, positivist discourse of power. This has led to a view that all we can ever really write about with authority and accuracy is ourselves, or at best, adopt an interpretivist/hermeneutic perspective that makes a virtue out of relativism and subjectivity, transforming ethnography into autobiography (or texts which Geertz (1988:84–97) criticises as 'I-Witnessing', 'confessional' and 'author-saturated').

Claims such as these raise all sorts of debates about whether there are such things as 'facts' or 'objective truths' in anthropology, or whether the postmodern critique has succeeded in collapsing the

Cartesian subject/object dualism. The reflexive/interpretive approach, however, seems to overlook two key issues. First, it presumes a degree of self-awareness and insight on the part of ethnographers that has rarely been empirically verified. Why should we assume that anthropologists – who are supposedly professional 'strangers', trained in the study of others – are any good at knowing or representing themselves? Second, even if anthropologists do possess such a critical self-awareness, this doesn't even begin to address the problem of how they should locate themselves in their texts. If fieldwork, as I have argued, spills over into influences, events and contexts that precede or come after the conventionally classified 'fieldwork period', how should anthropologists accommodate these considerations in their reflexive accounts? In short, what are the limits of reflexivity, and who is this 'anthropological self' that appears in the self-reflexive text?

To write about ourselves necessarily requires a degree of 'objectification of the self'. In doing this, however, anthropologists tend to adopt certain literary tropes and conventions. What emerges is invariably a contrived, edited, or 'authored' version of the anthropological self. Very often, what presents itself as a 'self-reflexive' and personalised account of fieldwork turns out to be nothing of the sort – as Crapanzano (1986) illustrates in the case of Geertz's famous essay, 'Deep Play: Notes on the Balinese Cockfight' (1973a). In this case, and many others, the 'reflexivity' trope actually conceals more than it reveals about the author and, worse, prevents the reader from getting close to the subjects of ethnographic enquiry.

PRE-FIELDWORK FIELDWORK, OR, FORMATIVE EXPERIENCES OF A WOULD-BE ANTHROPOLOGIST

Having highlighted some of the many pitfalls anthropologists face when writing ourselves into our narratives, let me turn to my own fieldwork in Italy and, at the risk of falling into the elephant traps described above, narrate some of the critical 'pre-fieldwork' fieldwork experiences that shaped it.

Like many other anthropologists that I know, university was where I first 'discovered' anthropology, and I eventually signed up for as many courses (or 'modules' as they were called) as possible on my joint anthropology and human geography degree. As well as stimulating enthusiasm for the subject and for learning, writing essays, reading anthropological texts and being exposed to social science concepts encouraged in me a more critical consciousness – or what C. Wright Mills and others have called the 'sociological imagination'. Thinking about other lifestyles, other conceptual systems and other moral universes made me think critically about my own society and convinced me that I wanted to live abroad. Much of my motivation to live abroad stemmed from a sense of dissatisfaction with the English way of life, particularly the narrow-minded conservatism ('ethnocentrism' seemed a fitting term) and the emotionally repressed character of the English (as I saw it). Durkheim's idea of anomie and Marx's concept of alienation also struck a chord with how I saw life in the UK. While I wanted to do fieldwork, however, I never wanted to become a professional anthropologist, and still less an 'academic' which was typically equated with ivory towers and navel-gazing disassociation from the 'real world'. However, intellectual curiosity made me determined to continue in higher education.

Death in Venice: learning about 'face-to-face' society in Burano

My first fieldwork experience, as a final year undergraduate, involved two months on the island of Burano in the lagoon of Venice, living with a highly conservative and Catholic family with whom I did not get on at all well. The island was tiny (you could walk from one end to the other in about five minutes) and had a population of barely 5,000 people. Although the stay was short, I learned a great deal, particularly the point made by Goffman (1959), Bailey (1971) and others that 'face-to-face' communities, as anthropologists often call them, are also 'back-to-back' communities. The goldfish-bowl nature of this society was striking. 'Surveillance society' is often equated with modernity and bureaucracy or Foucault's (1977; 1991) notion of the rise of the modern forms of disciplinary power, but it seemed to be very much a feature of small community life. In Burano, everyone seemed to know everything about everyone else's business, which led to interesting practices of secrecy, competition and gossip. The whole repertoire of informal means of social control that I had read about in ethnographic monographs was displayed before me, like so many textbook examples. I was a prize asset for the family I stayed with: paraded as a status symbol and a companion for their son. The problem was that I was extremely uncomfortable in the role they expected of me (child; associate member of the family; not allowed to befriend other families; expected to be compliant; to eat everything given to me). I experienced what it must be like to be a young teenager in a traditional Italian family – and I felt suffocated: so much so that I had to escape periodically to the anonymity of Venice. However, all of this helped to bring home to me the importance of age, sex, status, kinship, neighbourhood and common interest which, having just read Lowie's *Social Organisation* (1948), gave me a framework for thinking about the way societies may be patterned.

Marshall Plan or Martial Law? Becoming a doctoral student

I started as a postgraduate at Sussex University with a three-year grant from the Social Science Research Council (SSRC). My research proposal was to study the meeting of two global but local ideological systems, communism and Catholicism. I was inspired by David Kertzer (1980) and William Christian (1972), among others, but also critical of them for not, as I arrogantly argued, going far enough into the questions of meaning, identity and ideology that their work raised.

The Chair of SSRC's anthropology committee had suggested I do fieldwork in Poland and explore the role of the Catholic Church in a communist state. I was all ready to make a reconnaissance visit and had set up accommodation at Cracow University when the famous strike broke out in the Lenin shipyard in Gdansk. Soon afterwards, Solidarnosc was created, General Jaruselski declared martial law, Warsaw Pact tanks took up position on Poland's borders – and it looked like another Prague Spring was about to happen. That put paid to my plans to do fieldwork in Poland. Thereafter, I and my supervisors agreed that Italy would be a safer bet.

Caught in the crossfire: the witches of Kemptown

One of the most memorable aspects of undergraduate 'training' at Sussex at that time was that there was, in effect, no training and no MA programme. Like most anthropology students in those days, I had been steeped in the mystique of fieldwork: fieldwork seemed to be a magic rite of passage that separated ignorant neophytes from 'trained, professional anthropologists'. I read all the relevant literature on Italy and European anthropology from which I learned that European societies were characterised by family and neighbourhood, political patronage, class conflict and codes of honour and shame – that is, apparently not all that different from the Latin American peasantry from where many of the early insights and models used in European ethnography had been appropriated. I also learned, as most anthropology postgraduates are taught, that while other disciplines may arrive at similar conclusions, only anthropology – with its 'fine-grained' thick descriptions based on 'intensive' fieldwork – can unravel the complexity and nuances of social life and cultural practice as they are actually lived. Yet despite the supposed centrality of fieldwork to the discipline – now elevated to an almost mythical status – there was no course in methods. My 'preparation for fieldwork' consisted instead of one brief seminar with David Pocock who passed on pearls of wisdom from Evans-Pritchard, such as 'bring a stool with you', 'don't get food on your fieldnotes' and 'get interested in whatever interests the natives'. Other anthropologists of my generation and older have told me that they received even less 'methods training' than this. The attitude at that time was that 'you can't teach anthropological methods' because fieldwork was, by definition, too idiosyncratic, personal, unique and unpredictable: you either had the knack and could cope, or you did not. With retrospect, I came to see that this stance reflected more than simply 'bad practice'; it was also an expression of a kind of macho ethos quite prevalent among many older male anthropologists. Fieldwork was 'character building'; the tough crucible in which good ethnographers were forged. Like army training, therefore, it should not be made too easy.

Fieldwork training was therefore largely a DIY job. One of my MA papers was an ethnographic study of the local Labour Party. This involved five months' fieldwork looking at cliques, factions and informal groupings in a local ward of the Kemptown constituency Labour Party which, as I was to discover, was dominated by the Trotskyist Militant Tendency. Taking on the role of 'ever-available, naive anthropologist', I ended up being 'recruited' into the Militant Tendency – and being invited to attend covert caucusing meetings, ostensibly billed as 'Marxism Reading Groups'. Then early in 1982 the Labour Party National Executive launched an official inquiry into Militant to judge whether its members were Trotskyist 'entrists' or not (my informants denounced this as a 'witch-hunt against true socialists'). The result was that the situation grew increasingly tense and polarised and my own position became untenable.

The lessons I took from that experience was that fieldwork can be highly sensitive and politically fraught – and it raised all sorts of ethical dilemmas about confidentiality and how my data could be manipulated if I had allowed it to be publicised (which I determined I wouldn't do). However, it also helped make sense of the material I was reading about ethnicity and identity – especially Barth (1969), Abner Cohen (1974), R. D. Grillo (1980) and A. L. Epstein (1972) – and about how boundaries are constructed and maintained, and the role that ideology plays in group organisation. Indeed, the configuration of 'Ideology, Organisation and Identity' gave me the framework for organising my PhD several years later.

Panic in Perugia: where are 'the natives'?

I chose to go to the central Italian city of Perugia because I was interested in urban anthropology. At that time this was seen as the cutting edge of new anthropological research: we had a plethora of studies of small, remote, rural communities but little was known about life in European cities. I also chose Perugia because it was a regional capital (and university city) located in the heart of the communist 'Red Belt' zone (or *cintura rossa*) and because I wanted to study middle-class intellectuals rather than peasants and workers. The paradox of Italy, and its fascination for me, was that here was a major Catholic country (seat of the Vatican; a country where the presence of the Church can be witnessed in every walk of life), and yet postwar Italy had witnessed the rise of the largest Communist Party in the Western world – and the creation of a political and social system in which party politics extended into every aspect of life (aptly named the '*partitocrazia*' or 'Party-ocracy').

I arrived in Perugia early one morning in July 1982, after a long train journey from London, with suitcase and portable typewriter in hand and a heavy rucksack on my back. My first fieldwork experience, after finding a room to rent, was one of panic: I rapidly began to feel that ethnography in a big city was simply not possible. The problem was how to get to know the locals? I also had no idea how one decided what constituted the proper 'unit of study'. Nothing had prepared me for this. In Perugia I only seemed to meet foreigners. Shortage of time and funding had meant that I had no language tuition prior to my arrival, so my first act was to enrol on an intensive language course at Perugia's so-called 'University for Foreigners', which made the problem worse. I was living in a house with Palestinian and Syrian students, then later shared a flat with two German medical students both of whom had come to Italy because, unlike German medical schools, there are no limits on student numbers. My predicament was not how to understand the natives' point of view, but how to meet the natives in the first place.

The broad focus of my research, initially, was religion and politics. I therefore soon got involved with a local evangelical Catholic group called '*Comunione e Liberazione*' (Communion and Liberation, known colloquially as 'CL') and a local branch of the Italian Communist Party (the *Partito Comunista Italiano* or PCI). I also did some language tuition – more as a way of meeting locals than out of financial need. After four months I began to get to grips with the language. After six months I had made several Italian friends from the different circles I was frequenting, but I wasn't really trusted in either the Catholic or communist camps: I'd sit in on section meetings (highly ritualistic events which would last for hours); I was tolerated, humoured, but kept at a distance (apart from one dyed-in-the-wool Stalinist (or *filosovietico* in PCI-speak) who took me under his wing). I remember reading Kertzer's account of how local PCI members in Bologna tried to make sense of this outsider in their midst: many were convinced that he was an agent working for the CIA. My informants saw me not as a spy so much as a student and teenager (that is, not an adult) – and therefore not somebody to be taken seriously. I also attended Church outings: lots of sitting in circles clapping hands and singing upbeat songs of praise accompanied by acoustic guitars and smiling, bearded open-toe sandalled priests. My Catholic informants saw me as a candidate for missionary work and a target for conversion.

At this point, three friends were my mentors and guides to Italian culture and society. My closest friend, Primo Tenca, was 36 years old: a quiet but perceptive man of extraordinary generosity. He was also secretary of the local PCI section that I had started to frequent. His life-history was a paradigm of the social and economic changes that have shaped Umbria since the 1940s. He had been born into a poor sharecropper family in a small village close to the border of the two Umbrian provinces of Perugia and Terni, but at the age of 13, as industrialisation was transforming the region, signalling the end of the old Mezzadria sharecropping system and way of life, he had been apprenticed to a jeweller. He also become politicised and joined the PCI at about that time. From that apprenticeship he had become a qualified goldsmith and had finally been able to buy his own small workshop in a quiet street near the centre of Perugia. This was a regular meeting point for friends and party activists, as well as customers. I would often spend days in his company, sitting in the corner of his shop asking questions or simply reading while he crafted a piece of metal into a gold wedding ring or brooch. Primo was my first real window onto PCI culture and Umbrian society and, thanks to his generosity and patience, I became increasingly involved in his world.

At about the same time I met a 38-year-old teacher who worked in Perugia's prison and was looking for someone to exchange Italian for English conversation. Her life story opened my eyes to another key episode in modern Italian history. She had been a student leader and political activist in one of the many far-left groups that had formed during the 'movement of 1968': a self-styled *sessantottina* (or 'generation of '68'), as they were classified, but now somewhat disillusioned with politics. She had recently returned to Perugia after living in Turin for ten years and was finding the city's life provincial and claustrophobic. My hunger for knowledge about Italy was matched by her fascination with the 'Anglo-Saxon world' as she called it.

My third key informant at this point was a young law student; a devout Catholic woman from Puglia whose social world hinged largely around local church-based activities (which, in the university were dominated by members of Communion and Liberation), and the Scout movement. In Italy, I should point out, Baden-Powell's invention had been transformed into a predominantly Church-dominated organisation run by the clergy. Our friendship seemed to be based upon fascination of the 'Other'. Innocenza's ambitions were to complete her law degree, find a husband and return to Manfredonia (where she would work in her father's law practice), and start a family. She greatly admired Pope John Paul II and enthusiastically supported his conservative social policy, including the Church's teaching on chastity, marriage and the family. In the eyes of my PCI friends, she was an anachronism, part of the 'lost generation' and somewhat typical of the 'southern mentality'. My *sessantottini* friends were less phlegmatic: for them, she was the embodiment of Catholic hypocrisy, neo-fascist values and the 'disgusting petit-bourgeois mentality' that they had been struggling against since the 1960s. Her attitude towards them was equally hostile. Knowing that my research concerned the relationship between Catholicism and communism, she proposed one day that I meet her local parish priest who had been one of those rare '*Catto-Comunisti*' ('Catholic-communist') priests who, in the more radical climate of the early 1970s, had sought to unite Marxist discourse with the social teaching of the Catholic Church. As I later discovered, the visit had another agenda: she and the priest wanted to save me from the atheistic demons, and I was being vetted as potential marriage material. That relationship ended shortly afterwards. A year later Innocenza was married to one of her fellow Scout activists and had a child.

Caught red-handed: a 'flyposting by night' encounter

It was the summer of 1983, the Italian general elections were looming and 'CL' were taking a lead in promoting the electoral campaign of the ruling Christian Democrat Party. One night, after midnight, I received a knock on my door (by then I had moved in with an Italian-Argentinian – a self-professed Peronist) in the heart of the neighbourhood (or '*quartiere*') that I wanted to study. Outside were five members of the local PCI section executive, all ready to embark on a flyposting raid. So off we went, sticking '*VOTA COMUNISTA*' posters on top of '*VOTA DC*' posters around the city (which CL had placed). At about 1 a.m., under a bridge near the old Roman aqueduct, with my communist friends and paste brush in hand, we were suddenly interrupted by a group of about nine CL members from my local group. They recognised me immediately and asked what on earth was I doing with 'those communists'? As one CL activist once described my PCI friends to me, these people represented '*la parte patologica della società italiana*' ('the pathological part of Italian society').

Having to explain myself to them inevitably heralded a rift in my relationship with the CL informants. As a result I gained 'trust' from my communist informants (and later friends), but I had burnt my boats with the Catholics. What I'd read about the Catholic/communist division in Italian society being merely superficial was simply untrue: maybe in small towns and villages or among casual supporters, but in this city they represented two very different cultures and lived in utterly different conceptual universes (what one Italian sociologist called '*il mondo cattolico e il mondo comunista*'). That said, however, the Catholic Church still monopolised the rites of passage. It is therefore difficult for Italians – even sons of atheistic communists – to escape its influence. That point was borne out in numerous ways, for example in the comment of one of my communist friends' father, a peasant from the countryside outside of Perugia. One day at their home, having been invited to the important Sunday lunch ritual, this 70-year-old man turned to me and asked whether I was a Catholic. No, I replied. A Protestant? No, I replied. Jew? No, I said, adding that I hadn't been baptised into any religion. '*Come una bestia!*' ('like an animal!') was his answer to that. I didn't take it as an insult, more as an insight into the Catholic worldview and its tradition conflation of personhood (and civilisation) with Christianity (I later learned that '*un cristiano*' was, in fact, a colloquialism for 'person').

The flyposting encounter was a turning point in my relationship with the local party activists. Subsequent incidents confirmed this: for example, being party to more intimate conversations where comments were made to the effect '*poi parlare: lui é compagno*' ('it's okay, you can speak candidly: he's a comrade'): being introduced to strangers with pride as *un compagno inglese* ('comrade from England') or as *un compagno dalla mia sezione* ('comrade from my Party section'). Then a year later I was invited to attend a training course for party cadres at the PCI's national school outside Rome. Following this, I was finally given access to the Federation archives, which were full of confidential files on individuals, living and dead (many of which had been gathered during the Stalinist period).

My role as participant-observer studying the PCI also involved some critical and compromising decisions. One of the canons of fieldwork is that we get close to our subjects; we get interested and involved in the things they get up to, and where appropriate, we do what they do and go where they go. The focus of my research was local party activists (or '*militante di base*'), party intellectuals and the regional party elite: I wanted to answer the question, what does it 'mean' to be communist in Italy? In order to find out, I had to join in, which is what I did: I joined the PCI and became a local party activist. That entailed getting up at 7 a.m. on Sundays to sell the party newspaper, *L'Unitá*, going on rallies, marches and joining demonstrations, attending meetings several times a week and helping with the annual local Party festivals, the *Festa del Unitá*. But becoming a PCI member was no easy decision: I felt moral pressure to join, but there were pressures and good reasons not to. In 1983, the arms race was intensifying; President Reagan was making manichean speeches about the communist menace and the 'Evil Empire'; the Second Cold War was here. In Britain, Mrs Thatcher and her government had declared war on socialism – and were gearing up to take on the 'enemy within'. Joining a communist party – even a foreign one – was 'not a good career move', so my supervisors told me afterwards. It was also quite unusual for a British person to belong to an Italian political party (although another Briton was at that time Secretary of the university branch of the PCI). However, from the point of view of the local PCI activists, my being a member of the British Labour Party (which, in 1983, still regarded itself as socialist) meant that I was already classified as 'close kin', politically speaking. Becoming a Party member was a symbolic step and recognition of affinity on both sides.

SOLVING THE RIDDLES OF CULTURE: POST-FIELDWORK FIELDWORK

I returned to Britain after 18 months in Perugia and started writing up. After six months, my grant ran out so I found a part-time job teaching geography to undergraduates. I finished my thesis in September 1985 and was awarded a DPhil a few months later (Shore 1985). Shortly afterwards I was offered a job as a political research officer. However, on the day of my interview I also received an offer of a teaching job at Perugia University. After agonising over the decision, I opted for Italy (and the more unpredictable future), and that was the start of an even more intensive year of 'fieldwork'.

What lessons, then, did I take away from this first major fieldwork encounter? Fieldwork in Italy confirmed many of the things I had read in European ethnographies, but it also taught me to treat those ethnographies with caution. I was struck by the poverty of theory as far as urban anthropology was concerned, little of which had equipped me for this kind of encounter. During the entire 18 months in Perugia, I never once heard anyone mention 'honour' or 'shame', which were still regarded in the anthropological literature (if increasingly doubtfully), as the key values of Mediterranean society (Peristiany 1966). I had read Sydel Silverman's (1975) ethnography of an Umbrian hill town with its vivid description of a fundamental and pervasive ideology called *civiltà*[3] – but nowhere did I find any evidence of it in the Umbrian capital or its surrounding villages. Kertzer's (1980) thesis that people joined the Communist Party primarily for social rather than ideological reasons was plainly incorrect as far as my Perugian data went. Further, I realised that this 'social-ideological' dichotomy was itself analytically flawed as a model, and reflected more the 'ideological' concerns and assumptions of American political scientists than empirical reality.

I also gained a sense of the importance of history (or rather, constructions of history) for defining boundaries and identities, and the uses of history for establishing legitimacy and authority. History is central to the way most groups construct their identity (Tonkin et al. 1989). However, for PCI activists in Umbria it was much more than this: history was a process, a charter for action, an integral part of their consciousness; it was the *raison d'être* for the party and for their role within it. It gave flesh and meaning to their social world and was a unifying phenomenon that many regarded as a possession (in the sense of 'Party history', the PCI's 'historical legacy', its martyrs and heroes and its celebrated 'patrimony of ideas'). In a curious sense, I felt as though I had lived through twentieth-century Italian history vicariously: the rise of fascism, the March on Rome (which started in Perugia), the Second World War, the partisans and the Resistance, the Cold War, the Hungarian uprising, the May '68 movement, the Red Brigades, 45 years of Christian Democracy and corruption scandals, and now the PCI leadership's pronunciation of the end of the historical phase begun by the Bolshevik revolution – these events were recounted to me in vivid detail. Fieldwork made me aware not only of their importance, but also of how they were experienced and interpreted by different party members: how the past was forever present in the present.

One of the research topics I focused on during fieldwork was political socialisation. To answer the question 'what does it mean to be communist in Italy?', it was necessary to explore when people had joined the PCI and what had motivated that decision. I therefore conducted dozens of taped interviews asking party members to recount their life-history from a political perspective. These interviews provided one of the core themes of my thesis and subsequent book: the concept of 'political generations' (or, *pace* Shakespeare, the 'Seven Ages of Communist Man'). From these narratives I was able to plot out systematically the key political events and changing cultural influences that had shaped successive cohorts of party recruits over the years. This helped me to understand current internal party divisions as each 'generation' had formative experiences, political icons and ideological conceptions peculiar to itself. It brought home to me forcefully the extent to which the 'personal' is political, and vice versa.

Trying to follow the intricacies of Italian politics also alerted me to the workings of power and how political systems really function. I had come to Italy full of what were, with hindsight, rather naive, 'functionalist' and typically 'English' views about the nature of politics and government. Living in Italy changed all that. Here I discovered a country where corruption existed almost everywhere: where political parties had effectively 'clientised' millions of employees and officials (cf. Shore 1990:66–70; 1989): where ministers and government bureaucrats were systematically creaming off the assets of state-owned enterprises and illegally awarding lucrative contracts to friends and cronies; where the secret services were behind terrorist bombings of trains and stations and the kidnapping and murder of the Christian Democrat leader, Aldo Moro; where leading members of the cabinet, the judiciary, the media and the armed forces belonged to the secret 'P2' masonic lodge which planned to take over the state; where senior ministers were in collusion with the Mafia; where Giuglio Andreotti, veteran leader of the Christian Democrat Party (DC) and five times Italy's prime minister was under investigation for conspiracy to murder, and where most of his cabinet colleagues – including former Prime Ministers Bettino Craxi and Silvio Berlusconi, and one-third of Italy's former elected parliamentarians – had been formally charged with corruption. Learning these details, sometimes experiencing them directly, was an eye-opening experience – particularly for someone who came from a country where the height of political scandal was for a politician to commit adultery or lie to the House of Commons.

Fieldwork also made me aware of what Gramsci and others might call the 'tide' of historical events and the forces of history – macro-events that really did have a profound effect on the world in which my informants lived: Israel's invasion of Lebanon (an important event in my neighbourhood given the relatively large concentration of Palestinians and Arabs); the invasion of Afghanistan and the clampdown on trade unionism in Poland (which had prevented me from doing fieldwork there); the 'Star Wars' initiative; the decline of Soviet hegemony. The year I spent observing section meetings was the year the PCI finally broke with the Soviet Union: the next few years witnessed not only the rise and fall of 'Eurocommunism' but the collapse of the Berlin Wall, and the break-up of much of Eastern Europe.

I learned from my Italian friends to see how these events were interconnected. What was significant about 'the natives' point of view' – and their at times overdeveloped sense of history and politics – was that it was impossible to analyse in a detached, objective way: their worldview contained an explicit critique of my own society. It was impossible to analyse their 'folk model' without questioning my own. And I found that this did challenge many of my previously held views. In Italy, for example, I learned that what was politically unthinkable, in British terms, was commonplace. I remember the collapse of the Banco Ambriosiono in 1982 and the macabre death of the banker, Roberto Calvi, whose corpse was found hanging under Blackfriars Bridge in London, with stones in his pockets. The coroner's verdict of 'suicide' was greeted with incredulity and derision in Italy. I was appalled. My communist friends took the news phlegmatically. 'Did you think there weren't any freemasons in London?' one of them asked me, mockingly.

Italian communists were adept at situating local relations in a global context – something which anthropologists and 'globalisation' theorists have only recently caught onto. In PCI-speak this was called '*inquadrare*', or 'framing' the problem. Every PCI rally, opening congress speech, or leader's address would be prefaced with a lengthy exegesis of the 'current historical phase', including the latest perspectives on the international situation and its ramifications for national, regional and local affairs. I once commented on this with irony, saying 'you can't talk about the state of Perugia's pavements without relating it to the price of the dollar or the latest policy decision of the Pentagon'. 'That's right,' came the reply, 'when America sneezes, Italy catches a cold.' In their view, American foreign policy and international capitalism were what set the agenda. 'Why do you think the Christian Democrats have remained in power for so long? And who do you think is behind these outrages?', I was asked, rhetorically. In 1983, most observers would probably have dismissed these suggestions as political paranoia. Ten years later, with the Cold War over and the ruling Christian Democrat Party defunct, their accuracy had been all but vindicated.

PCI activists in Perugia took a keen interest in Margaret Thatcher's trade union reforms, particularly the miners' strike, recognising long before I did the significance of its outcome for the future of the European left. The concept of 'growth without jobs' and the 'two-thirds society' – with one-third of the population consisting of a permanent underclass of the long-term unemployed – were being debated by rank-and-file party members even before these ideas had entered mainstream political debate in Britain. The PCI's Marxist vocabulary provided a framework for thinking and talking about economic and political processes which could be both simple, yet also extremely sophisticated. According to my PCI friends, the struggle between capital and labour moved like a pendulum: during the boom years of the 1950s and 1960s, economic growth, combined with pressure from an increasingly well-organised left, had forced the capitalist class to relinquish some of its privileges, allowing a far greater percentage of the population to share in the nation's wealth. Now the pendulum was swinging back: growth had slowed, profit margins were being squeezed and capitalism was once again trying to strengthen its hold over power and privilege. That was how they interpreted the shift to the right in Britain, and *Inghilterra*, they thought, was an experiment which would be repeated by other governments.

As well as introducing me to details of British colonial history that I had never been taught at school (one of my friends being an expert on naval history), I also learned to see my own society as they saw it. When Italian friends came to stay with me in London, or for a holiday in Britain (as they did quite often), I played the role of host and tourist guide, often visiting places which, as a Londoner, I had never found time to see before. My 'grand tour' of London would invariably include (at their request) visits to the British Museum and Parliament, some typical English pubs, London's parks and galleries, the old Stock Exchange, and, as a matter of course, Karl Marx's grave in Highgate Cemetery. I was also made aware of some of the more banal cultural practices which, for my Italian friends at least, were a source of some hilarity. 'Why do you wash up your dirty plates and cups with a toilet brush?', one asked me, 'And why do you English never rinse the plates after you've washed them?' Another commented on the English obsession with domestic pets and fire regulations asking why, 'if you're so terrified about fire, are your houses stuffed with thick carpets and curtains?' She also thought the combination of thick-pile carpeting and domestic pets disgusting and unhygienic ('Imagine, all those dog hairs stuck in the carpet fibres: you English must love housemites and cockroaches!'). Another quizzed me about houses in England: 'Why do none have shutters on the windows: aren't they cold in winter?'; 'If England is a country of "individualists", why are the houses so uniform?' I had no easy answers for these and the many other questions about English cultural practices: most of them I had never even considered.

CONCLUSION

These, then, were just some of the 'critical' experiences that shaped my fieldwork in Italy. Although my academic interests have shifted to other areas and I no longer work on Italian issues, that encounter with Italian culture was neither ephemeral nor superficial. Much of my understanding of the meaning of history and politics and its rival interpretations was forged in Italy: my culinary tastes and habits have been indelibly stamped by Italian cooking; the (secular) godparents of my children are Italians, and I still speak to and see my Italian friends whenever possible, although some relationships have inevitably grown more distant over time. My view of Italian society has also changed: after working in an Italian state university and being exposed to the patronage and clientelism that only later acquired the descriptive neologism of '*tangentopoli*' (or 'kick-back city'), I realised why my friends often described employment practices and work relations as the 'dark side' of the Italian experience (see Shore 1989).

My fieldwork in Italy has not so much 'ended' as 'metamorphosed'. The years I spent in Perugia were only a small part of that fieldwork, and many other factors influenced that transposition from personal experience to objectified text which we call ethnography. Rethinking the concept of 'fieldwork' also has implications for the way we conceptualise 'the field'. The traditional idea of a clearly bounded space or people has become increasingly problematic: 'the field' proper is a fluid, loosely connected set of relations, sites, events, actors, agents and experiences from which, and onto which, anthropologists try to impose some kind of conceptual order. Moreover, anthropologists are not like 'detached' scientists studying the behaviour of rats from outside a glass cage; we are positioned subjects within those fields and should therefore be 'objects' of anthropological enquiry as well. Writing more candid, subjective and reflexive accounts of 'what really happened during fieldwork' does at least help to render this more apparent – to ourselves as well as our readers.

Fieldwork stories, however, offer only part of the picture. Equally important in determining what anthropologists write about and how they write it, as I have tried to illustrate, are the pre-fieldwork and post-fieldwork encounters. Subjecting these to critical 'reflexive' scrutiny and situating them in our texts is possibly more problematic than the conventional 'reflexivity genre' of anthropological writing, but it is also more fruitful. Indeed, 'reflexivity' proper comes in many different guises (and disguises) and serves different agendas. I think it is useful only when it serves as a springboard for examining wider questions of theory and method, or for analysing the way anthropological knowledge is produced. A more critical and analytical reflexivity should also oblige anthropology to turn its professional gaze towards the broader context in which its research practices are embedded, including the institutional setting in which anthropology takes place (particularly the conditions of its existence within the university system). That, in turn, means taking into our analytical framework more 'reflexive' questions such as 'who funds the research?' and 'what influences are brought to bear on that research by government policy, research council funding strategies, publishers' demands, commercial constraints and other key agencies that shape academic and intellectual life?' When anthropologists do address these issues it is usually retrospectively; that is, half a century of more after the event – as in the case of Asad's (1973) analysis of the relationship between anthropology and colonialism. All of these institutional/contextual factors play a role in shaping our discipline and the kinds of fieldwork that we do. In short, they too are part of the critical pre- and post-fieldwork fieldwork experiences that we so often forget to include in our anthropological frame of analysis.

NOTES

1. This point has been raised by many other anthropologists too. See Shore and Wright (1997) for a discussion of ways of 'reconceptualising the field' in anthropology.
2. Anthropologists who have worked in small communities usually also become quite sensitive to the politics of reputation management – which some would argue are endemic in most close-knit or face-to-face societies.
3. As Silverman (1975:2) describes it, *civiltá* refers to a cluster of popular beliefs and values about 'civilisation' and a 'civilised way of life'. Perugian informants were perplexed by this description and pointed out *civiltá* was an archaic and somewhat aristocratic idiom that they had rarely encountered – except in the speeches of Mussolini.

REFERENCES

Armstrong, James (1994) 'Goy in the Promised Land; or, Murphy's Law and the Outcome of Fieldwork', in DeVita, P. R. (ed.) *The Naked Anthropologists: Tales From Around the World*, California: Wadsworth Publishing Co., pp. xii–xxiv.

Asad, Talal (1973) *Anthropology and the Colonial Encounter*, London: Ithaca.

Bailey F. G. (ed.) (1971) *Gifts and Poison: The Politics of Reputation*, Oxford: Blackwell.

Barth, Fredrik (ed.) (1969) *Ethnic Groups and Boundaries*, Boston: Little, Brown and Co.

Christian, William (1972) *Person and God in a Spanish Valley*, New York: Seminar Press.

Clifford, J. and Marcus, G. (1986) *Writing Culture: The Poetics and Politics of Ethnography*, Berkeley, CA and London: University of California Press.

Crapanzano, Vincent (1986) 'Hermes Dilemma: The Masking of Subversion in Ethnographic Description', in J. Clifford and G. Marcus (1986), pp. 51–76.

Cohen, Abner (1974) *Two Dimensional Man: An essay on the anthropology of power and symbolism in complex society*, London: Routledge and Kegan Paul.

Cohen, Anthony P. (1993) 'Post-Fieldwork Fieldwork', *Journal of Anthropological Research*, Winter: 1–23.

Du Bouley, Juliet (1974) *Portrait of a Greek Mountain Community*, Oxford: Clarendon.

Epstein, A. L. (1972) *Ethos and Identity: Three Studies in Ethnicity*, London: Tavistock.

Foucault, Michel (1977) *Discipline and Punish*, Harmondsworth: Penguin.

Foucault, Michel (1991) 'Governmentality' in G. Burchell et al. (eds) *The Foucault Effect: Studies in Governmentality*, London: Harvester Wheatsheaf.

Fox, Richard (1991) 'Introduction: Working in the Present', in Fox R. (ed.) *Recapturing Anthropology: Working in the Present*, Santa Fe, NM: School of American Research Press/University of Washington Press, pp. 1–16.

Geertz, Clifford (1973) *The Interpretation of Cultures*, New York: Basic Books.

Geertz, Clifford (1973a) 'Deep Play: Notes on the Balinese Cockfight', in Geertz (1973).

Geertz, Clifford (1988) *Works and Lives: The Anthropologist as Author*, Cambridge: Polity Press.

Goffman, Irving (1959) *The Presentation of Self in Everyday Life*, New York: Doubleday.

Grillo, Ralph (1980) *'Nation' and 'State' in Europe. Anthropological Perspectives*, London: Athlone Press.

Grimshaw, Anna and Hart, Keith (1995) 'The Rise and Fall of Scientific Anthropology', in Ahmed, A. and Shore, C. (eds) *The Future of Anthropology: Its Relevance to the Contemporary World*, London: Athlone Press, pp. 46–65.

Kertzer, David (1980) *Comrades and Christians: Religion and Political Struggle in Communist Italy*, Cambridge: Cambridge University Press.

Kohn, Tamara (1995) 'She came out of the field and into my home', in Cohen, A. P. and Rapport, N. (eds) *Questions of Consciousness*, London: Routledge, pp. 41–59.

Lewis, Ioan (1985) *Social Anthropology in Perspective*, Cambridge: Cambridge University Press.

Lowie, Robert (1948) *Social Organisation*, New York: Holt, Rinehart and Winston.

Polier, Nicholas and Roseberry, William (1989) 'Tristes tropes: post-modern anthropologists encounter the other and discover themselves', *Economy and Society*, 18 (2):245–64.

Peristiany J. G. (1966) *Honour and Shame. The Values of Mediterranean Society*, Chicago: Chicago University Press.

Rabinow, Paul (1977) *Reflections on Fieldwork in Morocco*, Berkeley, CA: University of California Press.

Shore, Cris (1985) 'Organisation, Ideology, Identity. A Social Anthropology of Italian Communism', unpublished DPhil thesis, Sussex University.

Shore, Cris (1989) 'Patronage and Bureaucracy in Complex Societies: Social Rules and Social Relations in an Italian University', *Journal of the Anthropology Society of Oxford*, XX (1): 56–73.

Shore, Cris (1990) *Italian Communism: The Escape From Leninism. An Anthropological Approach*, London: Pluto Press.

Shore, Cris and Wright, Sue (1997) 'Policy, a new field of anthropology', in Shore, C. and Wright, S. (eds) *Anthropology of Policy: Critical Perspectives on Government and Power*, London: Routledge.

Silverman, Sydel (1975) *Three Bells of Civilization: The Life of an Italian Hill Town*, New York: Columbia University Press.

Stocking, George (1992) *The Ethnographer's Magic and Other Essays in the History of Anthropology*, Madison, WI: University of Wisconsin Press.

Tonkin, Elizabeth, McDonald, Maryon and Chapman, Malcolm (eds) (1989) *History and Ethnicity*, London: Routledge.

Tyler, Stephen (1986) 'Post-Modern Ethnography: From Document of the Occult to Occult Document', in Clifford and Marcus (1986), pp. 122–40.

Wolf, Eric (1982) *Europe and the People Without History*, Berkeley, CA: University of California Press

2 LOCATION AND RELOCATION: HOME, 'THE FIELD' AND ANTHROPOLOGICAL ETHICS (SYLHET, BANGLADESH)

Katy Gardner

As all pre-doctoral students of anthropology are aware, fieldwork – usually in some far-flung location – is the discipline's centrepiece, the ultimate transformative experience through which they must pass if they wish to call themselves anthropologists. Amongst my fellow neophytes at the LSE (1986) we approached this great experience with a mixture of romantic expectation, heroic self-image and utter terror. What we were all certain of (and none of our ritual masters disabused us) was that whatever happened, so long as we stayed the course, we would come home fully-fledged members of an elite group, having completed the greatest academic initiation rite of all. Our training reiterated this. Exposed in our pre-fieldwork seminar to the tales of various post-fieldwork initiates, we gradually understood that within anthropological epistemology, knowledge is generated and validated through individual experience. Far more important than learning the appropriate interview or survey methods was the act of doing, of being there and letting the local culture permeate our individual boundaries. To put it crudely, we were taught that the way in which anthropologists learn is by going native. In this chapter, I wish to update this romanticised version of anthropological transformation. What I hope to show is that rather than the relationship between our transformative fieldwork experiences and the texts which result from them being lineal (what we learnt during fieldwork determines what we write), it is in fact far more complex. This is partly because what and how we learn is endlessly influenced by our personal locations and identities, which themselves change over time. It is also because anthropological learning is not bounded by the temporal boundaries of being in

49

the so-called 'field'. To support my argument, I shall draw from recent debates within postmodern critique and feminism.

Anthropology has of course had its fair share of postmodern castigation. One area in which it has been criticised is the claim of so-called 'objective generalisation', or what Jonathan Spencer calls 'ethnographic naturalism' (1989:153–4). This confers authority on the anthropologist by suppressing the historical specificity of the ethnographic experience; the experiential nature of data collection, which is so much part of private anthropological lore, is left out of the public end-product. If we are to meet the postmodern challenge, we therefore need to describe the historical specificity not only of the people who we write about, but also of ourselves. (For a classic exploration of these issues see Clifford and Marcus 1986.)

The importance of locating the author in her or his text has now been accepted by most anthropologists and is virtually a literary convention. This is an important step forward, but often simplifies the anthropologist's own identity and position. I, for example, would present myself in terms of various characteristics (female, white, middle-class, etc.) from which particular assumptions would then be made. The premise of this is first that race, gender and class are more influential in my positioning than other characteristics (the list could include any number of random characteristics, such as age, educational background, sexual preference, voting behaviour, and so on), and secondly, that these characteristics can be read like a map reference (if I am white, middle-class, etc., I am therefore located between points Y to Z). This issue has been particularly discussed and analysed within feminism. Before considering my own fieldwork experiences, I shall therefore make a short digression, and outline some of the key issues which have arisen from postmodern feminism.

Q4

FEMINISM AND THE REVISED POLITICS OF LOCATION

Just as being there confers authority on anthropologists, so is personal experience the key to feminism. Rather than separating this from their writings however, feminists have always celebrated the personal. For many first-wave feminists, women's experience of oppression was core to their subsequent consciousness raising, and the basis of their new political identities. As Visweswaran has commented, feminism 'locates the self in the experience of oppression in order to liberate it' (1988:29, quoted in Mackey 1991:5). Crucially, these experiences were understood as shared. Belief in the universal subordination of women and their international sisterhood was thus, for many, a central tenet of their feminism.

This assumption of shared experience was to lead to increasing self-criticism and fragmentation as over the 1970s and 1980s the movement became aware of difference, at first in terms of sexuality, and then later in terms of race (Mackey 1991:2). The critiques tore at the heart of Western feminism; it was accused of generalising from what were essentially white and middle-class experiences, of ethnocentricity, and of appropriating black women's voices. Some argued that white feminism's attempts to represent Southern experiences helped reproduce neo-imperial power relations, and thus colluded and collided with orientalism (Mohanty 1988).

As a response to these criticisms, by the late 1980s what has been called the 'politics of location' became increasingly influential. This involves the recognition that everyone writes from specific locations, and that the ways in which we learn and experience have particular temporal and spatial dimensions (Probyn 1990). Thus, there can be no homogeneous, universal feminism, for all individuals have different locations. Here then, it is difference which is highlighted; such ideas have obviously been heavily influenced by postmodernism.[1] The politics of location have however led to a new set of problems, for they often assume an automatic relationship between experience, identity and political position. Identity politics are thus in continual danger of essentialism: individuals are assumed to represent particular groups whose cultural and political locations are presented as fixed and bounded. They are also often heavily prescriptive, dictating who can and cannot speak. They may be used to argue, for example, that dialogue can only take place between women who share the same identity. In other instances, they may mean that the authentic voices of women of colour are heard, whilst those of white privilege are silent. Both of these situations are highly problematic (see Mohanty 1990; Mackey 1991).

Identity is of course far more complicated, for individuals possess many identities, none of which are fixed. This, plus the rigid oppositions implied by what Mackey has called a dynamic of 'authentic voice/privileged silence' (Mackey 1991:5), have led to a revised politics of location within feminist thought. In this, location is understood as multiple and always changing; individuals are placed on shifting ground where no single identity is possible. Instead, personhood is endlessly fragmented; we have permeable boundaries and are endlessly transformed in each new interaction. The positions one speaks from are located in and contingent upon specific contexts. The researcher and her informants are both changed, and agents of change.

ANTHROPOLOGY AND THE REVISED POLITICS OF LOCATION

Similar issues exist in anthropology. To avoid essentialising an anthropologist's identity, or assuming its coherence, we can demonstrate how our own personhood is multiple and at times contradictory. It is continually changed, both through the ethnographic experience and in interactions before and after. Anthropology is thus not simply about specific discoveries (although these may be important), but is more processual, building upon lessons which begin before the intrepid neophyte arrives in the field, and long after she or he has left. As Nancy Lindisfarne has written:

The logic of a post modernist position means that the sense and authority of any new ethnics [*sic*] of anthropology can only be derived from self-critical analysis of the micro-politics of all interactions ... not only those of the anthropologist in the field, but those at home before and after fieldwork. (1994:6)

Rather than presenting fieldwork as bounded in space and time (my experience in the field), and the anthropologist as trapped in the static identities of before and after her or his transformative experience, it is therefore more constructive to understand it as an ongoing process, for each anthropologist's relationship with her or his experience is continually changing, just as we continually change as individuals.

In what follows, I am therefore not going to describe an experience or revelation which changed the course of my fieldwork. Indeed, when I tried to recall such an experience I realised that there was no single revelation which made everything clear. Instead, I wish to focus on how my understandings of my fieldwork have been altered through my own changing political, personal and intellectual locations. I also hope to indicate the relationship of these to various ethical problems.

I have been particularly forced to consider these issues both because of the nature of my research topic, and because I have produced three distinct texts from my ethnographic experiences: a book of short stories, my PhD dissertation and its completely rewritten incarnation as an academic text. Each of these represents a different phase in my relationship with the place where I did my fieldwork (a village in north-eastern Bangladesh) and a different type of understanding. This process of change encompasses not only my first and longest stint in the so-called field but also subsequent periods in Britain, and the various times which I have returned. In what follows, I shall explore these issues by discussing two particular issues: purdah and migration. The latter, in particular, has embroiled me in various ethical questions, which have invariably been closely interwoven with my particular positioning.

FIELDWORK: A BACKGROUND

I did my fieldwork in a village in Greater Sylhet, Bangladesh, over 1987 and 1988. As an undergraduate, I had taught English to a Bangladeshi woman and like most of the British Bengali population, which then included around 100,000 people, she came from Sylhet. I therefore decided to live in a village in Sylhet and study the social and cultural effects of prolonged overseas migration. In preparation for my visit, I worked for about six months as a volunteer for a community centre in Spitalfields, London, and through various contacts met a number of British Bengali community leaders. During this stage, I was located within the represented British race relations, in which the British Bengali community is discussed largely in terms of racism, victimisation and exploitation. I did not problematise my own position, or my research plans, but assumed that eventually, I might use my experience in a positive way in the UK.

Once in Bangladesh, I was lucky to find a place to work in relatively swiftly. Talukpur is a small village of 70 households in Nobiganj, one of the areas in Sylhet which send migrants to Britain, to the Middle East, and increasingly, to the US. I moved into a large homestead consisting of four patrilineal households. Two of these were in Newcastle, and the others were present. My adoptive family were of high status, but owned only a small amount of land. As I quickly realised, most households with migrant members were considerably more prosperous than those without. There was therefore increasing economic polarisation in the village, for the main way in which people could move upwards was to work abroad and earn foreign revenue. In this sense, migration is a highly valued economic resource to which people continually struggle to gain access. In general, only those with a certain amount of capital or pre-existing contacts on the migrational network ever got the opportunity to go abroad.

I stayed in Talukpur with my adoptive family, to whom I became an honorary daughter, for 15 months, give or take breaks in Dhaka. Occasionally I went to Sylhet Town and got drunk with the local VSO workers. In the village, I worked from an agenda of questions, which generated more issues to follow up. I carried these in my head from household to household, repeating the same questions, memorising people's answers, and at the end of the day, writing down everything I had heard. I tried to visit all the households in the village regularly, but inevitably spent more time with some than others. Originally I took notes during my conversations, but for reasons I'll discuss below, soon decided that it would be better to be less explicit and simply memorised what people told me. Later on in my fieldwork, I tape-recorded interviews and the oral histories of particularly keen informants. I also wrote down everything that happened. My supervisor had assured me this should fill about six pages of A4 a day. On some days, I confess, I could only muster half a side.

As with all anthropologists, my relationships with my informants changed and developed over the period I was in Talukpur. Throughout my stay I was in a variety of roles, often all at the same time. I was, in varying degrees with different people and according to context, a white European (thus of high status), a possible spy, a young, unmarried woman, and a fictive daughter, sister, niece, auntie, etc. Certainly, my role as researcher, which was never very clear to start off with, either to me or to my hosts, became more and more blurred. If I was to be part of things and truly accepted, I increasingly sensed that I could not also be a researcher; it felt too detached and too hierarchical. Possibly I tried too hard to conform and thus swept the research carpet from beneath my own feet, for how could a village daughter go around inter-viewing people and writing things down in her book? Towards the end of my stay I became increasingly paralysed in my formal research role. Needless to say, it was in the very last months of fieldwork, when I stopped asking direct questions and everyone knew that I was leaving, that I learnt most. These roles have changed further on my return visits to Talukpur in 1990, 1993 and 1994. I am still an honorary daughter in name, but am obviously less close to everyday affairs than I was; possibly I may be drifting into being a patron; certainly, I am no longer a PhD student taking notes and carrying around a tape recorder.

PURDAH AND GENDER

The approaches of Northern feminists to social institutions such as purdah (the veil) have been central targets in the critiques of first-wave feminism I alluded to earlier. Southern intellectuals such as Chandra Mohanty have argued that Western feminists make ethnocentric assumptions about purdah, and in presenting all veiled women as passive victims, uncritically promote their own culture and position (Mohanty 1988). In some instances this has certainly been the case, but it is also important not to essentialise the positions of so-called Western feminists. Personally, I do not have any one view on purdah. Gender relations in Talukpur have continued to confuse, challenge and infuriate me. In this section I intend to chart the relationship between my experiences in and out of Talukpur, my changing locations and my ethnographic understandings.

When I arrived in Talukpur I had a fairly straightforward view of purdah. Veiling was the tool of patriarchy; it subordinated women, whilst giving vast privilege and power to men. This attitude was partly the result of previous personal experiences travelling in the Middle East and South Asia. It was also strongly reinforced by the existing literature on gender relations in Bangladesh. Until recently, this has almost wholly stressed Bangladeshi women's oppression, what has been called a 'litany of grim statistics' (Arthur and McNicholl 1978). As Sarah White has argued, the majority of this literature has been funded by aid agencies whose priorities hugely influenced the subsequent research agendas. The problem of Bangladeshi women is thus given centre stage, whilst other aspects of gender relations are ignored (White 1991). I must confess that before I arrived in Talukpur I assumed that the women I met would envy me. If possible, my presence might awaken their latent feminist consciousnesses.

During my relationship with Talukpur, much has confirmed my original assumptions. If I chose to work in a Muslim society partly out of a desire to venture into the heart of the alien other there was, and is, much in Bangladesh to satisfy my voyeurism. One of the first things I was told in the village was: 'Women's heaven is at their husband's feet.' Women are denied entry to all male domains (the mosque, the market, the fields and the village council). In Muslim law, one man carries the weight of two women, who are legal minors. Within rural areas they also have very few rights, despite the official laws made in Dhaka. Their children, for example, belong to their husband's lineage and if they are divorced, they often lose them. Women are defined by, and dependent upon, men. Many of the married women who I knew were nameless, labelled through their relationships to men, as so-and-so's wife, or so-and-so's mother. At marriage, the bride's behaviour wholly symbolises the passive submissiveness of stereotypical rural Bangladeshi women. Heavily veiled, she does not even walk without the assistance of her female kin. Later, in her in-laws' house, relatives and neighbours flock to view her, lifting up her veiled face to take a look and then commenting on her relative fairness. In none of the marriages which I observed did brides have any choice in who they were married to. Ostensibly, at least, they present their husbands as their future masters, whom they must obey and please in all respects.

I could continue my description. This chapter is not, however, about the subordination of Bangladeshi women, but of how my understandings of local gender relations changed over time. Let me start with my physical experience in Talukpur, for one of the most important ways in which I learnt about Bangladeshi women was through my own bodily transformation. As I was quickly to learn, constructions of Bengali femininity are inscribed first and foremost on the body. This has been described by other female outsiders working in rural Bangladesh. Kotalova, for example, writes of how her body was examined by local women when she first arrived in her fieldwork village and then transformed by them as they taught her their own bodily codes. As she suggests, the ambiguity she posed presented a challenge to their own identities and assumptions about femininity (Kotalova 1993:28–34; see also Ram 1991).

My first lessons in Talukpur were also of a physical nature (see Gardner 1991). For my hosts, it was imperative that I become a proper woman, and I was rigorously instructed as to how this might be possible. My hair was oiled and tied tightly back, glass bangles pushed onto my wrists, my fingernails stained with henna; most importantly, I was instructed to change my *shalwar kameez* (baggy trousers and tunic, worn by young girls) for a sari. From now on, my head should always be covered, especially when the *azan* (the call to prayer) was sounded from the mosque, or in front of elders. I should walk slowly, rather than stride or run, and sit only in certain positions. I should be smooth, neat, and demure: this was a sign of acceptable femininity. In contrast are women who lack the appearances of control. Their hair is loose and wild-looking; their saris are not properly tied, and may even hang open; their heads are uncovered, and they talk too much, a standard criticism of unpopular women. Those that truly stand at the margins of civilised society may even wander, with uncovered heads, into male domains. These mad women (*pagoli beti*) are the subject of much amusement and horror in Talukpur, and when I did not conform – when my hair was too frizzy, my sari rode up over my ankles, or I walked into male domains with an uncovered head – I was jokingly accused of being a *pagoli*, or a small child, that is, I could not be a fully formed adult for I had not yet been socialised into the correct social codes. This changed over time, for as I stayed longer in Talukpur, I began to get away with what I assumed to be deviant behaviour. In the hottest months, for example, I stopped wearing a blouse inside the boundaries of the homestead, concealing my chest with my sari, but leaving my back bare. To my surprise, no one objected, perhaps because I had already satisfied my hosts that I had mastered the basics.

There were other ways in which I learnt about Talukpur through my body. I developed, for example, a taste for *pan*, which women chew endlessly. When hungry, I found myself craving rice. This is of tremendous symbolic value in rural Bangladesh; by imbuing the substance of the locality through its food people are said to become linked to their homelands. It is only local rice, fruit, fish, and so forth which are seen as truly nourishing and tasty. My appetite for rice, then, indicated that I was slowly picking up the bodily mannerisms of the people I was with. In comparison with the often tense way in which we in Britain deport ourselves, and our notions of the need to exercise and use our bodies, these express more a state of relaxed indifference. I found too, that my use of Bengali, again reflecting the linguistic mannerisms of local people, was seeping into my personality. In

Bengali I am more abrupt, more extrovert and more assertive. In short, I began to feel my whole identity change.

These bodily experiences were not, however, simply a matter of learning a physical role and playing it like an actor, whilst remaining unchanged inside. Instead, personhood and physical experience are more intimately connected. As Marriott and others have argued with reference to South Asia, bodily substance and cultural behaviour are inseparable; each is a realisation of the other (Marriott 1976, cited in Kotalova 1993:64). Thus, just as wearing a suit might make some people feel more formal and restrained, after an initial period of awkwardness, my sari, oiled hair and bangles made me feel more demure and feminine. Slowly too, and perhaps partly because of this, I began to feel the first inklings of *sharam* (shame). This is a powerful cultural norm in Talukpur, and whilst not exclusively experienced by women, is largely associated with them. *Sharam* carries meanings of modesty, shame and embarrassment, according to different contexts. Whilst it is often caused by external events (such as a man unexpectedly appearing, before a woman has a chance to cover herself), it is also an internal state for women, for their bodies are constructed as inherently impure. All signs of fertility and sexuality must therefore be covered and hidden.

These feelings were not caused simply by wearing a sari, but also the result of being constantly reminded to cover my head in male company, and of being with women who would rush to hide themselves as soon as a strange man appeared. (Only certain categories of men evoke women's hasty removal from the scene. Landless labourers, beggars, or men too young to marry do not merit such behaviour.) Such is the power of socialisation that after about six months in the village I instinctively began to pull my sari over my head in front of male elders. Increasingly too, the three-mile walk over the fields to the nearest road, which at first I had insisted on doing alone, became more and more difficult due to my growing awareness of just how extraordinary my behaviour was. Even when chaperoned, I took to covering my face entirely with my umbrella. On one memorable occasion, I forgot to take this with me, and then had to walk through the male domain of shops and tea stalls to reach my destination. That I was horribly aware of the censoring eyes of village men as I hurried past indicated to me not only the power of social sanctions, but also the permeability of my own cultural boundaries. In another instance, I amazed one of my VSO friends in Sylhet. Sitting and chatting in his bungalow one evening, we were interrupted by the arrival of his Bangladeshi counterpart. As the employee of a progressive NGO, with experience of working with foreigners, there was no reason why I should feel compromised by this young man's presence. Instinctively, however, I ran into an adjoining room before he could see me, much to the hilarity of the VSO.

What this told me, of course, was how quickly my identity and position could change in different cultural locations. I was, and remained, a non-Muslim woman from Britain, with strong feminist sympathies. But instead of this automatically fixing me in a particular position, the ground on which I stood became increasingly slippery. Rather than possessing a homogeneous self, I was fragmented: drinking whisky one moment, and in my confused state rushing into self-imposed and inappropriate purdah the next. I am not trying to claim that my sense of *sharam* was the same as other women in Talukpur. All foreigners in rural Bangladesh attract large crowds, often composed entirely of men. Being stared at is not a particularly pleasant experience, and day after day becomes extremely tiresome. My tendency to cover up whenever I left the village, and was thus once more a stranger, was partly a practical solution to the problem of being foreign, rather than being a woman.

Combined with this, I manipulated purdah and used it to my own advantage, often in ways which would not be possible for insiders. It was, for example, the perfect excuse not to talk to the frequent male visitors who would come to our homestead, request to meet me, and then subject me to endless questions, often ending in an oppressive attempt to convert me to Islam. It meant too, that I could demand male help, when in other contexts I would have to manage alone. Errands could be run by my younger brothers, for example, and there would always be somebody to help me carry my bags on my trip to the road. If one has dependable male kin and is from a wealthy household, purdah can feel very comfortable. From this viewpoint, based on my own experience and observations of the women around me, women in Talukpur are not simply the victims of a harsh representative cultural code. Instead, I began to see how some might gain certain benefits from purdah, and the advantages of being relieved of the burdens of freedom.

Combined with this were the attitudes of the women themselves. Contrary to my expectations, they did not envy me; some even seemed to pity me. None embraced feminism in the way I had naively anticipated, although many expressed dissidence indirectly – 'We're of lower status than men, but more intelligent'). All confirmed that men were indeed hierarchically placed above them and possessed more power, for this was the rule of Allah. Rather than seeing purdah as a form of subordination, they mostly presented it to me as a central part of Islam, which was for many the most meaningful thing in their lives. To question this meant questioning God. I am wary of generalising, but suggest that most seemed to accept the outward, public presentation of gender, whilst working in a variety of contradictory and even subversive ways within the framework. As Kotalova (1993) has argued, we need to distinguish between the encompassing structure of social codes and the ways in which, within the encompassment, dissidence and negotiation take place.

Indeed, as time progressed in Talukpur, the tables turned and I began to find that rather than village women envying me for my freedom, I envied some of them for their security. For a young woman, in an economically comfortable and loving family, life is in many ways easier than for young women in Britain. None of the women I knew had to face the insecurity and uncertainty of forging their own destiny. They may have had little independence, but it is the family rather than the individual which is celebrated in Talukpur, and as time progressed I began to wonder if they may not in fact be right. Increasingly, whilst perhaps not intellectually acknowledging it, I began to emotionally appreciate why a culture which constructs women as in need of male care and incapable of independent decisions may not be wholly negative for those who have male support. Certainly, our British insistence upon individual freedom and choice began to appear almost as strange to me as it did to my hosts, whose statements such as 'In your country everyone is separate from their families, and alone' I found hard to refute.

My new understandings were not, however, static. They have changed over time and according to where I am, and my writing has reflected this. For example, my book of semi-fictional stories, some of which I had already drafted when I left Talukpur at the end of 1988, probably reflects my most subjective and emotional response to this apparent transformation. However, the story did not end there. I continue to reassess my view of gender relations in Bangladesh. Each time I return, I see things differently. This is partly because my personal boundaries have been rebuilt. As people told me in the last weeks of my fieldwork, 'Now you're going to make yourself into an English woman again' ('*tumi ekhan Inreji beti banabe*'). I no longer feel like an anthropologist in Talukpur, and thus am less tolerant of purdah. I feel freer to argue with my ex-informants. Since I now only visit for very short periods, empathy is also less crucial in maintaining the web of relationships and legitimacy on which the continuation of my work originally depended.

More importantly, I have witnessed changes in the lives of women that I have known for seven years. As their lives unfold, my own interpretations of gender relationships in Talukpur shift. Indeed, the male support I was reassured of in 1987/88 seems an increasing myth. Perhaps most influential in this change in my perceptions has been the fate of my closest friend in Talukpur, whose husband, when I first met her, was working in Saudi Arabia in 1987. She had been married about three years, but had only been with her husband for one month of this, and was childless. She was staying in her father's home, awaiting her husband's return. By the end of my fieldwork it was becoming clear that there were problems in the marriage. The husband had stopped sending letters, and relations with her in-laws were distinctly chilly. Since then, the marriage has completely broken down. When, after five years away the husband returned, he declared that in his absence my friend had grown too old for him, and he intended to take a second wife. She would have to find another path in life. For the vast majority of rural women in Bangladesh, marriage and children are the only respectable path in life. My friend, now approaching 30, and childless, was ruined. To make matters worse, the head of her lineage has not so far allowed her to divorce her husband, which would enable her to take him to court to claim her *kabin* (money set aside for the wife as marriage settlement in case of divorce) and give her some financial independence, for it might damage the status of the lineage, and hence his own political career.

My outrage at this has reconfirmed many of the views which I held at the beginning of my fieldwork, and which began to slip throughout it. But paradoxically, my writings on gender have moved in the opposite direction. When I wrote my thesis in 1989, I did so mainly within the subordination of Bangladeshi women discourse. More recently, and coincidental with my discoveries of postmodern critiques of anthropology and Western feminism, I have been at pains not to label Talukpuri women as victims and have attempted to be more detached, to stress the processual aspects of gender, the diversity of experiences and meanings, and so on. My subjective feelings and my academic analysis are therefore in constant danger of contradiction. This personal problem is echoed by the recent debates within feminism which I discussed earlier.

As Chandra Mohanty (1988) has argued, to interpret South Asian gender relations and institutions such as purdah solely in terms of inequality and female subordination negates individual agency, homogenises women's experiences and presents them as a single and wholly victimised category. If female informants refuse to recognise social and religious codes as subordinating, if they appear to accept their social roles, and indeed, make strenuous efforts to persuade the anthropologist herself to follow their codes of behaviour (from covering her head to Islamic conversion and even an arranged marriage), then surely describing them as subordinate (and hence in a state of mystification) is deeply patronising? Rather than attempting to locate overall, systemic structures of subordination, as embodied by notions such as patriarchy, a postmodern approach would focus more intently upon individual experience and agency, and the specificities of local and historical context. It may also problematise the writer's own position, even to the extent of denying her a voice because she is not an authentic representative of the correct group, and thus, since there is no universal female experience and no universal women, she cannot speak for the experiences of the other.

These postmodern correctives are undoubtedly important in their recognition of diversity, and in pointing out the dangers of homogenisation, and the hegemonic categorisations of many Western-centric discourses. The difficulty with them is, however, that in other senses they are inherently depoliticising, for if everything is fragmented into endless diversity, it becomes virtually impossible to talk of structures of inequality, exploitation or subordination. As others have pointed out, too, the assertion that so-called white or Western women have no right to speak about experiences other than those ascribed to their particular characteristics, is highly dangerous and likely to lead to the wholesale collapse of feminism. Writers such as Nicholson have therefore suggested that there must be postmodern stopping points and that gender is one such point (1990). Each time I return to Bangladesh I become more convinced that this is correct.

The challenge facing all anthropologists who hope to produce politically committed work is thus how to write about diversity without apparently undercutting such stopping points. To insist that there is no single view of Bangladeshi women, or of purdah, for they, like the anthropologists, are continually in different roles, and continually changing, is a start. Likewise, we need to recognise the various levels at which social norms work, from encompassment (Kotalova 1993) to the manipulation and negotiation of overarching structures. This explains the various contradictory messages which I received: women should be covered, modest, passive and so on, but there were also many ways in which they implicitly deviate, and work within the rules. In other words, it is not enough merely to see the explicit and external symbols of women's subordination (the veil, official laws and norms) and assume that there is no more to learn. These lessons do not mean that we are unable to take up political positions, but we must also recognise that these positions are not fixed. The challenge, then, is to continually revaluate our own locations, with renewed sensitivities to both our own fluidity and that of the people we discuss.

MIGRATION

If writing about gender has inescapable political implications, writing about migration is even more fraught. The migration of Sylhetis to Britain involves very real political issues, both in terms of state policy – immigration law, council housing allocations and so forth, and the racism many face in Britain. The ways in which they are portrayed are thus particularly important and, understandably, many British Bengalis are tired of their misrepresentation by outsiders. I have therefore had to step through a potential minefield. Before discussing these directly I would like to describe some ways in which I experienced this sensitivity and how my anthropological reactions to it have both been informed by my pre-existing locations, and in turn influenced those locations.

Many people in London warned me that Sylheti migration would prove to be a difficult research topic. I understood these warnings in two ways: as evidence of how relevant my research was, and as a methodological challenge. Since I did not plan an exposé of illegal immigration, I also thought that such advice was probably alarmist. It is worth noting that at this stage I was solely interested in the problems posed by my forthcoming experience. I did not consider I might face problems after completing my initiation; I had only considered the ethical problems of data collection – for example, does one pay one's informants? – not the ethical problems of what I would do with my data (and certainly not whether I should be collecting it in the first place.

Following the advice of my supervisor, the first thing I did in Talukpur was a rough survey of the village, visiting every household and asking who lived there, who was related to whom, and so forth. If I was told that members were abroad I enquired where they were and how long they had been gone. These questions, in addition to the basic fact of my presence, led most people to conclude that I must be a British High Commission spy. Given that the BHC's immigration officials regularly carry out what are termed 'village visits', in which they arrive unannounced in migrant villages to check up on the details of immigration cases, this was entirely rational. That I was asking 'when did you last see your father?'-style questions also did not help; neither did my regular absences in Dhaka.

I realised that I was generally thought to be a spy very gradually, by the vague hostility of some people, and the advice of my adoptive family, who remained loyal to me throughout. Some men directly challenged me, asking what I was doing and whether I could prove it. I had a letter from my university stating I was a student, but I doubt this did much good since no one could read English and the letter could have been easily forged. I said that I was student, learning from them so that I could write a book; I am sure that for most people this cut little ice. These fears indicated to me the central importance of migration to people's lives and the extent to which they fear, often for good reasons, the immigration authorities. Moreover, as I will explain shortly, it has deeply affected both the content of my work and my relationship to it.

So far, my unease was confined to my relationship with people in Talukpur and the need to convince them I was not a spy. After about four months, however, things developed in a different direction. On a visit to Sylhet Town, I was introduced to one of the many social workers, teachers and community activists who visit the area from Britain. When I explained to him what I was doing he became very aggressive, telling me that whatever I ended up writing could only harm the Sylheti community in Britain; even if I did not intend it, my work would be used by racists against Sylhetis. He also insisted that all research was a waste of time and self-indulgent.

This experience was extremely distressing to me, but it did raise my awareness of the sensitivity of my work and the need to write extremely carefully. Perhaps then, it was no bad thing. Likewise, it was no bad thing that I embarked on my fieldwork so naively. If I had started it knowing what I know now, I doubt if I would ever have collected the basic survey data I did in my first months. This data has been invaluable, and since originally a sister from my household accompanied me on the visits to give me introductions and assist me in understanding the Sylheti dialect, it later proved to be largely correct. To this extent, ignorance was bliss.

Migration has therefore caused a variety of methodological, ethical and theoretical problems in my work, which I have attempted to solve in different ways. My reactions to these problems and solutions are closely intertwined with my particular political and geographical positions, which, as with the issue of gender, have been continually fluid. Methodologically, the local sensitivity to discussing migration meant that I changed direction. I stopped using my notebook, and generally disguised the extent to which I was recording things. I also spent more time than I had to with women, for they were far more welcoming, and less threatened by my presence. Centrally, I stopped asking about migration and only followed lines of enquiry which I thought would prove that I really was studying Sylheti culture. I therefore learnt about Islam, about local saints (*pir*), and healing, which people were happy to share with me. I was terrified of enquiring about land ownership and remittances, and most of my data in these areas was acquired indirectly (slipping the question into a conversation about something else, asking neighbours how much another household owned, and so forth). In some ways this has been an advantage rather than a weakness. Migration is all too often discussed only in terms of political economy. In my work, I have tried to understand its cultural and ideological meanings – these may have escaped me had I focused only on its economic effects.

However, whilst depoliticising my actual fieldwork, these methodological solutions have led to new problems in my post-fieldwork task of writing. Whilst conventional socioeconomic analyses of migration would be politically neutral in Bangladesh, writing about Islam has made my writing far more sensitive. Indeed, in the Bangladeshi context, Islam is the most loaded subject of all, even whilst in the village it was what people wished me to understand. Combined with this, my solution of concealing the extent to which I was recording data in Talukpur has meant that my fieldwork was highly top-down. It was not remotely participatory; although I told them I was writing a book, I did not tell my informants what I was planning to put in it. They have therefore had no forum for correcting me or putting alternative points of view. I realise that this is quite normal for anthropological research. Indeed, the theoretical complexity and rarified language of much anthropological discourse effectively excludes most informants, even if they could read English. It is an aspect of my work, however, which, whilst solving my fieldwork problems, leaves me profoundly uneasy in my role as a writer.

But as I was to realise on returning to Britain, far more pressing than the problem of anthropological exclusivity are the ethical issues posed by the main topic of my work: migration. Again, as a fieldworker, I had not clearly thought these through. These problems exist on several levels, the first of which is practical and the most easily solved. Clearly, I have had to censor my work. There are some things which I simply cannot write about, for if I did, it might directly endanger my informants and their families. This has remained unchanged in the various versions of my work. At the second level are issues which would not damage anyone directly, but might be used against the Sylheti community in Britain. The issues of housing and arranged marriages between London and Bangladeshi partners are two examples. Again, I have tried to exercise caution, although most of my data indicates the discrimination of British immigration law rather than the swindles of would-be migrants.

The next level is less easily solved. Overseas migration and its effects look different according to where one is geographically situated and politically allied. In the context of Britain, the British Bengali community is economically, culturally and politically subordinate. Especially in the context of recent political events in East London, anthropologists attempting to write about British Bengalis therefore have a responsibility to avoid replicating negative stereotypes or fuelling racist arguments. Before my fieldwork, I was located within these discourses.

In the Bangladeshi context, however, things look rather different. Migrant households are clearly the elite in rural Sylhet. They are often large landowners who employ many servants and live in large, pukka houses. If there is exploitation to be found – and Bangladeshi society is frequently characterised as one dominated by struggles for scarce resources and hence deep-rooted exploitation and corruption (see Hartman and Boyce 1983; Jansen 1987) – it is often perpetuated by successful migrants against non-migrants, who are almost invariably poorer and less politically powerful. From the outset, I have been interested in power relations. I did not want to write a thesis which dealt only with the cultural construction of personhood or symbolic codes. This has meant that issues of local differentiation and exploitation have been unavoidable for me. But should I write something which might reflect badly on an already beleaguered British ethnic minority?

The extent to which this question has worried me has largely depended upon my geographical position, and the audience which I am writing for. Whilst doing my fieldwork, I was located within the context of Bangladeshi, where it is virtually impossible to ignore poverty, inequality and the exploitation of the powerless by the elite. Indeed, nearly all of the anthropology of Bangladesh focuses upon these very issues (see, for example, Hartman and Boyce 1983; Arens and van Buerden 1977; Jansen 1987). The immediate academic product of my fieldwork – my thesis – thus dealt with the ways in which migration had changed local structures of power, and was key to the elite's manipulation of their cultural, economic and political dominance. This was written whilst emotionally at least I was still in Bangladesh. I had not fully engaged in anthropological literature beyond relevant South Asian ethnographies and the anthropology of migration. I was also secure in the knowledge that very few people would ever read my thesis. The immediate problems of representation were thus minimal. In my non-academic book, which I hoped might have a slightly larger readership, I chose not to write at all about migration to Britain in an attempt to dodge the issues. This was easy to do since the book's format was a collection of semi-factual stories which were based on my personal experiences.

While rewriting my thesis into a book, I have had to face these issues directly. I am also fully relocated in the British context. Although my book is primarily about Bangladesh, and not the Bengali community in Britain, my audience will be first and foremost British. I have therefore had to balance the need to be politically sensitive to the dangers of racist misinterpretation, my desire not to offend Sylheti readers, and my anthropological responsibility to be as true to my data as possible. I do not know if I have succeeded in these aims, and clearly it is impossible to please everyone. Various new approaches, which I only learnt of in returning from Talukpur have, however, helped me move some way forward.

First, I have not claimed that my version of Talukpur is either objective, or definite, and have used the now-routine technique of placing myself in the text as much as possible. Secondly, in stressing diversity and flexibility, rather than an overriding system one is less likely to homogenise or essentialise local cultures. Thirdly, I have found it more fruitful to focus upon individual agency and dynamism, rather than systems of exploitation. This leads me to my final point, which is that my relocation has been theoretical as well as physical, political and emotional. This, like my changing interpretations of gender, has been affected not only by doing fieldwork over a fixed time, but by returning to Talukpur.

My original analysis of migration was dominated by the available literature on development in Bangladesh (for an interesting discussion of this, see White 1992) and theories of migration produced in the 1980s (Meillassoux 1981, Castles et al. 1984, Cohen 1987). In both of these literatures, neo-Marxist notions of dependency and post-imperialism are key. Labour migration, like other flows which have taken place from colonised Southern countries to the North, must be understood first and foremost as a form of exploitation, in which value is extracted, and capital accumulated in the North, and the costs are borne by the South.

From these ideas, I argued that overseas migration from Sylhet could also be understood as a form of dependency. Along with other analyses of South Asian migration (for Pakistan see Ballard 1987; for Bangladesh, see Islam et al. 1987), my prognosis was therefore generally pessimistic. Migration increased local incomes for the elite, thus widening economic differentiation and making it harder than ever for low-income households to compete. It was also an extreme form of dependence. Not only was travel to *bidesh* (foreign countries) culturally constructed as virtually the only route for economic success and enterprise, but many households were financially dependent upon the remittances of absent members. I suggested that as in Britain at least, the British Bengali community became increasingly established, these would dry up, whilst in Bangladesh the villages' need for foreign revenue, like that of the national economy, would continue. Whilst benefiting individuals then, in the long term overseas migration would not benefit Sylhet.

Subsequently I have completely rethought this original analysis. This partly reflects the contemporary rejection of meta-narratives such as dependency theory within academic discourse, plus discussions of globalisation and the possibilities which these might provide for the anthropology of migration (see, for example, Hannerz 1992). Lastly, and most importantly, however, it reflects my return to Talukpur, which I revisited in 1993, after three years' absence. The first thing I noticed as I approached the village was the extent that, within this time, things had changed. The number of pukka houses had risen dramatically and many houses were currently being built. Combined with this, there was a new secondary school which had not existed in 1988. In 1997, what was the dirt track to the village is now a tarmac road and electricity has arrived. During my last visit I also noticed several agricultural changes: an increase in mechanisation amongst the larger landowners, and an increase in deep tubewells. My work has not focused upon agricultural development, so I should add that this observation is mostly impressionistic.

Most importantly, however, some of the households which had previously been in a pitifully poor state, had now significantly improved their economic position, often through the canny investment of remittances from the Middle East. Two of these landless households had new houses, and one had started a profitable tea business. This does not mean that overall poverty has decreased, for a large number of in-migrant landless households have been settled on government land by the local authorities. What it does suggest, however, is that within the social and political boundaries of the village, migration cannot be interpreted only as a cause of dependency and stagnation. Nor are migrants passive victims, tossed this way and that by the vagaries of the international labour market. Instead, they are highly dynamic. Again, I got a vivid sense of this in 1993, when I was told that since migration to Europe was difficult, new opportunities were now being sought in the Far East. Whilst dependency theory goes some way in explaining some of the structural conditions of Bangladeshi overseas migration, it therefore fails to indicate the dynamism and agency of migrants and their communities.

CONCLUSION

My changing relationship to my data, both during my fieldwork and after it, throws up a variety of more general issues. The first is that the epistemological assumption of anthropology, that we are simply objective conduits for our data, is clearly deeply flawed; for what and how we know is endlessly influenced by our various shifting locations. Rather than being passive conduits for the data, we are proactive in choosing what we learn and what we write. To suggest that anthropologists need not worry about political or ethical issues because their first duty is to academic truth therefore borders on irresponsibility. Instead, what we learn and what we write is unavoidably subjective, because we are all located in particular political positions. As I have also suggested, our learning does not end when we finish our fieldwork, and there is therefore no definitive account of it.

My account also indicates that the ethical issues encountered in the field often appear in a different light at home. Topics which are neutral in one context are therefore sometimes highly charged in another. This has been a particular issue for me because the people I am writing about are also living in Britain where their political and social position is very different. Combined with this, researching and writing are different acts and throw up different problems. For example, my learning about Islam was seen by my hosts as wholly positive. Publishing anthropological analyses of local Islam has, however, been fraught with political dangers, especially in post-Rushdie Britain, where much of the popular representation of Islam is virulently anti-Muslim. The much-publicised case of Taslima Nasreen, the Bangladeshi feminist writer, has also made commentary on gender relations within a Muslim village in Bangladesh highly sensitive.

Lastly, I would like to suggest that what we experience in the field has a direct bearing on our relationship to the anthropological endeavour and our reactions to postmodern critiques of it. The fears and suspicions which my presence evoked have also, I think, changed my relationship to my own writings and to anthropology. For a phase in my fieldwork I became acutely paranoid about what people thought of me. I was terrified that I would meet direct hostility, or even aggression, and would not be able to continue. More profoundly, the experience of being held in suspicion and not always being welcome is deeply unsettling. Discussing these issues with friends who did fieldwork with people who welcomed their enquiries, and who were vociferous in their desire to be written about, I realise that postmodern critiques of anthropology have fewer emotional resonances for them than for me.[2]

Perhaps the issues which I chose to study are particularly sensitive. Although painful, in the long term this has been no bad thing for it has forced me to confront some important ethical dilemmas, not only in the doing of fieldwork, but also in writing it up. What I have tried to describe in this chapter is how over time my dynamic relationship with Talukpur has affected me personally and intellectually. Likewise my various and changing positions and identities have altered the way in which I have approached the village, on both an intellectual and a personal level. These relationships are continually changing. Only in writing do they appear to become concrete and set; perhaps that is why it is such a challenging business.

NOTES

1. The fragmentations implied by postmodernism, and its insistence on endless difference have led some feminists to worry that it may lead to the movement's self-destruction (see Nicholson 1990; also di Leonardo 1991:24).
2. To a degree, my own qualms are unfounded, for what people were afraid of was that I would give to the British High Commission details which would lose them their immigration cases, not that my representations of them would be a form of cultural imperialism. Many people enjoyed telling me their stories, and today seem pleased when I show them my book.

REFERENCES

Arens, J. and Beurden, J. (1977) *Jhagrapur: Poor Peasants and Women in a Village in Bangladesh*, Birmingham: Third World Publications.
Arthur, W. and McNicholl, G. (1978) 'An analytical survey of population and development in Bangladesh', *Population and Development Review*, 4 (1): 23–80.

Ballard, R. (1987) 'The political economy of migration: Pakistan, Britain and the Middle East', in Eades, J. (ed.) *Migration, Workers and the Social Order*, London: Tavistock.

di Leonardo, N. (ed.) (1991) *Gender at the Crossroads of Knowledge: feminist anthropology in the post-modern era*, Berkeley, CA: University of California Press.

Castles, S. (with Booth, H. and Wallace, T.) (1984) *Here for Good: Western Europe's New Ethnic Minorities*, London: Pluto Press.

Clifford, J. and Marcus, G. (eds) (1986) *Writing Culture: The Poetics and Politics of Ethnography*, Berkeley, CA and London: University of California Press.

Cohen, R. (1987) *The New Helots: Migration and the International Division of Labour*, London: Gower.

Gardner, K. (1991) *Songs at the River's Edge: stories from a Bangladeshi village*, London: Virago (reprinted by Pluto Press, March 1997).

Gardner, K. (1995) *Global Migrants: travel and transformation in rural Bangladesh*, Oxford: Oxford University Press.

Hannerz, U. (1992) *Cultural Complexity: Studies in the Social Organisation of Meaning*, New York: Columbia University Press.

Hartman, B. and Boyce, J. (1983) *A Quiet Violence: View from a Bangladesh Village*, London: Zed Books.

Islam, H. et al. (1987) *Overseas Migration from Rural Bangladesh: a Micro-study*, Chittagong: Chittagong University.

Jansen, E. (1987) *Rural Bangladesh: Competition for Scarce Resources*, Dhaka: Dhaka University Press.

Kotalova, J. (1993) 'Belonging to Others: Cultural Constructions of Womanhood Among Muslims in a Village in Bangladesh', *Stockholm: Acta Universitatis Upsaliensis, Uppsala Studies in Cultural Anthropology*, 19.

Lindisfarne, N. (1994) 'Local voices and responsible anthropology', in Stolcke, V. (ed.) *Reassessing Anthropological Responsibility*, London: Routledge.

Mackey, E. (1991) 'Revisioning home work: feminism and the politics of voice and representation', MA term paper, University of Sussex.

Meillassoux, C. (1981) *Maidens, Meal and Money: Capitalism in the Domestic Economy*, Cambridge: Cambridge University Press.

Mohanty, C. (1988) 'Under Western eyes: feminist scholarship and colonial discourse', *Feminist Review*, 30: 61–88.

Murray, C. (1981) *Families Divided: The impact of labour migration in Lesotho*, Cambridge: Cambridge University Press.

Nicholson, L. (ed.) (1990) *Feminism/Post-modernism*, London: Routledge.

Probyn, E. (1990) 'Travels in the postmodern: making sense of the local', in Nicholson (1990) pp. 176–88.

Ram, K. (1991) *Mukkuvar Women: Gender, Hegemony and Capitalist Transformation in a South Indian Fishing Village*, London: Zed Books.

Sobhan, R. (1982) *The Crisis of External Dependence: The Political Economy of Foreign Aid to Bangladesh*, Dhaka: Dhaka University Press.

Spencer, J. (1989) 'Anthropology as a kind of writing', *Man: the Journal of the Royal Anthropological Institute*, 24 (1): 145–64.

Stephens, J. (1992) 'Feminist Fictions: a critique of the category non-Western women in feminist writings on India, in Guha, Ranajit (ed.) *Subaltern Studies VI*, Delhi: Oxford University Press, pp. 92–106.

White, S. (1991) *Arguing with the Crocodile: Gender and Class in Bangladesh*, London: Zed Books.

3   ON ETHNOGRAPHIC EXPERIENCE: FORMATIVE AND INFORMATIVE (NIAS, INDONESIA)

Andrew Beatty

INTRODUCTION

Much recent debate on the nature of ethnography centres on the question of experience: what goes into it, what can be extracted from it, how far it determines or merely frames knowledge, its specificity and privacy versus its generality and availability. The debate is never far from polemic. There are those for whom subjectivism is a term of abuse – think of Freeman on Margaret Mead – and those for whom it is almost a principle, and for whom the sin is to be unaware of what is methodologically unavoidable.

Renato Rosaldo (1993), for example, holds that what the ethnographer brings to the field in the way of personal experience determines the limits of his or her understanding. 'Has the writer of an ethnography on death suffered a serious personal loss?', he challenges. Until Rosaldo had direct experience of grief, he was unable to understand or accept Ilongot explanations of why they go headhunting: namely, to work off the 'rage' associated with bereavement. Understanding comes with empathy and empathy derives from common experience.

Rosaldo invites us, therefore, to be suspicious of ethnographic reports which do not incorporate some personal experience or demonstrate their empathy in a way similar to his own example; apparently, he would limit research to topics on which one is qualified not by study but by experience. This kind of argument, incidentally, rules out the possibility of imaginative literature, and of learning from being with and talking to others. In fact it practically rules out communication on all but the most superficial technical subjects.

But we have only to reflect on our experience to recognise the limitations of the argument. For while Rosaldo clearly has a point – that a young person cannot fully understand what it is to be old, nor a woman know what it is to be a man, and so on – we are, nevertheless, able to feel sympathy when someone impresses us with an experience which we have not shared; otherwise there would be no point in recounting it. We understand something of Macbeth's ambivalence without having killed a king ourselves, and of Ahab's quest for revenge without feeling compelled to go out and harpoon a whale. And these are characters conjured from words on a page written in a different era; they cannot be questioned about the meaning of their actions. The telling of an experience is, moreover, itself a kind of experience, albeit a vicarious one for the listener; it is not merely a conveying of information. Its purpose is to kindle common feeling – sympathy – in the listener so he or she can grasp the subjective meaning – by empathy – of the other's predicament. How successfully the message gets across depends on shared knowledge and, perhaps, experiences which are analogous rather than common. (Of course, strictly speaking, no experiences are common.)

So we do not need to have lived through the same things to write perceptively about other people. We can make sense of what they do and say based on knowledge of the other person's circumstances, character and the cultural context. Ethnographic empathy rarely requires us to penetrate the inner thoughts and feelings of others, merely to interpret their expression correctly and to respond appropriately, though, of course, our interpretation will implicitly refer to motive and intention. If we are, additionally, to draw on our own experience as an ethnographic tool or as a criterion of authenticity we must be explicit about the epistemological differences between common experiences within the same culture, those which are analogous across cultures, and those which are shared in the field, albeit by people of different backgrounds.

Rosaldo assumes his personal grief to be sufficiently similar to that of an Ilongot headhunter to qualify as a key to their motives. But cross-culturally, what counts as common experience is highly problematic. The relation of person to event may differ in respect to agency, interpretation, expression and perhaps even to inner state. Javanese villagers often say that all human beings laugh and cry (and perhaps go on to add that all religions are one). But when a Javanese laughs it may be out of embarrassment rather than mirth; and the inner state, the embarrassment, is constructed differently than would be the case in another culture: the pressure to dissimulate or transform or suppress the feeling, all these are different. So what remains in common? Since there is no experience unmediated by culture, there is no direct route to cross-cultural understanding through common experience, even of the basic things in life. We must always first reckon with the problem of translation.

Experiences shared in the field are in a different category again. As we are socialised into the host community and begin to incorporate its unspoken presuppositions, we share – as our hosts, too, recognise – common experience in ever more subtle and comprehensive ways. And we begin to make better sense of our hosts' independent experiences. This is partly through the acquisition of background knowledge and partly through the dialectical movement between sympathy and empathy. What is increasingly at issue is how this growing understanding, this gradual meeting of minds, should be represented in the writing of ethnography. Is it a prelude to the published analysis or should it form the substance of the analysis itself? Do we need to keep the scaffold in place when the building is finished; or have we only been building scaffold anyway?

Ordinarily we tend to make a distinction between the kind of knowledge that can be communicated and the kind that can only be learned through experience (wisdom, know-how, certain skills, tacit knowledge). The former, informational, kind of knowledge is more easily rendered into ethnography. The latter kind – which has to do with judgement, discrimination, psychological understanding and ways of getting on with people – shapes our field experience and the writing of ethnography, as do our personality, inclinations, and so on; it makes someone an ethnographer of a particular cast. How we acquire such ethnographic wisdom – supposing we do – is far from clear. Probably most anthropologists would find it hard to draw a sharp line between what they learned by talking to people and what they absorbed through taking part. The current emphasis on dialogue, with its ethnocentric bias towards the tête-à-tête, claims to put all its cards on the table but actually masks the essentially mysterious process of coming to understand (or thinking one understands) what is going on in another culture and the very real part played by personal, sometimes unarticulated, experience in this process.

Those who would leave the epistemological scaffold in place are therefore in a quandary. For it would require extraordinary powers of introspection, self-knowledge and minutely detailed second-by-second documentation to be able to convey the myriad impressions of fieldwork and to recognise the changing patterns and schemata of social life which we construct on the way to our official, but always provisional, version in the published ethnography.

To make a dent in this problem, I would like to distinguish in this chapter field experiences which are ethnographically informative from those which are, so to say, 'merely' formative. Some experiences, of course, are both. But the reflexivists tend to emphasise their identity, as if what we learn and how we learn it are the same thing; or, to put it differently, that we could not have learned what we know any other way. I would like to suggest that the provisional schemata we construct are crucial to our growing understanding without being necessarily of lasting interest. What makes our final versions worth printing is not that they are true and the rest are false, but that they are our best estimates. Since our knowledge is constructed collaboratively with our hosts, this works both ways. The kinds of things people tell us are, at the start, grossly simplified or otherwise incomprehensible; at the end, ideally, they barely disrupt the stream of normal life in all its complexity. Dialogue gives way to conversation.

Fieldwork accounts often blur this distinction between the formative and the informative. A good story demands that a turning point or breakthrough with the host community should lead to a general illumination. Clifford Geertz, who espouses this doctrine of the ethnographic epiphany, has given us an example in his account of the Balinese cockfight (1973). He and Hildred Geertz began to be recognised as persons in Balinese terms only after fleeing a police raid, making a public choice of going Balinese. Subsequently, cockfights revealed themselves in their full intricacy as texts, social dramas and so on.[1] Perhaps it is worth noting that this justly celebrated essay, like much of Geertz's work, appeared first in a journal for a general non-anthropological readership more likely to accept or even expect this conventional fusion of social acceptance and ethnographic insight. In fact, what the essay demonstrates is not that his dramatic entry into Balinese life enabled him to see cockfights as texts; rather, his initial *exclusion* enabled him to understand something about Balinese personhood. The raid may have helped Geertz become an insider; but it does not help us understand the cockfight. As a rhetorical ploy, the report of the raid may persuade us to accept the account as true by establishing the author's credentials, but it does not make it true.

The growth of ethnographic understanding, if not a smooth upward curve, is at least a staggered progression, or perhaps a series of false starts and returns. It is very far from being, as Geertz suggests elsewhere (1983:70), like seeing the point of a joke – a mystification of 'How Anthropologists Think' if ever there was one. Indeed, this analogy could hardly be less apt in its connotation of sudden and final understanding: for there are not really two ways of understanding a joke; nor better ways. Nor, I would argue, is fieldwork a succession of epiphanies, each revealing some transcendent truth. A closer parallel is the acquisition of a language: impossible to say at what point one begins to understand or control it; impossible, too, to specify how and when one acquires each new grammatical feature. The process is mysterious, though not for some quasi-mystical reason. The gradualism I am postulating contrasts, then, with an empiricism which would radically separate observer and observed, and with various kinds of subjectivism which would either merge them or would make knowledge dependent on communion or 'participation' in a mystical, almost Lévy-Bruhlian, sense. The implications for the writing of ethnography, which I will come to later, are different in each case.

The anthropologist's prior experience, no less than her personality and training, is bound to shape perception in the field; but it is not simply a rigid mould into which the data flow, nor an obstacle to recognising other modes of experience. This is because the ethnographer herself is not a finished product any more than what she purports to represent. The point about participant-observation is that the two terms are in unstable and everchanging relation. The quality of one's participation changes with the quality of one's observation, and vice versa. And the growth of understanding is as much a shedding of preconceptions and a refining of explanatory schemata as a series of revelations. We often come to realise what we know some time later.

But aside from the controversial theoretical question of how observer and observed are mutually fashioned, there are certain practical, more obvious limitations on knowledge imposed by the circumstances of fieldwork. The accidents and contingencies of where we live, who we happen to live with or talk to and so on make an enormous but largely unacknowledged difference to the final product. (This is quite literally the case: how often do Acknowledgements dutifully list every critic, influence and supporter back home while lumping informants together anonymously as 'too numerous to mention'?) It could be argued that these contingencies are evened out or transcended as we learn to see what is typical or repetitive as opposed to random. In any case, all such contingencies are themselves culturally embedded and it is of interest to see how *our* 'accidents' or coincidences are construed by our hosts in the field. To give an example, early on in my recent fieldwork in Banyuwangi, in eastern Java, I met a man who came to be one of my closest friends and teachers. As I understood it, our paths crossed in the following manner. I wanted to make a playpen for my daughter and needed a kind of bamboo that didn't splinter easily. I had seen a particular variety which had an attractive brown pattern referred to as its 'batik'. This bamboo is uncommon, but a man at the other end of the village was mentioned and I went to see him. Pak Sunar had the very thing and without any bargaining we made the deal. The playpen, which the villagers to my consternation called a cage, was hardly ever used and eventually was eaten by ants, but I had made an important acquaintance. Much later on, Pak Sunar quoted to me words his guru had told him 30 years ago: 'There will come a person, buying or selling: do not refuse him.' The buying or selling is meant, of course, metaphorically, as seeking some kind of answer or perhaps (though not in my case) offering one. Our accidental meeting had been foretold. And so it was Pak Sunar began to teach me about Javanese philosophy.

FIELD PERSONA AND POINT OF VIEW

Who we meet and how, and what to make of it: these accidents are often passed over in the writing of ethnography because, presumably, sooner or later in the field one connects up with the right people anyway; or, to take the opposite point of view, unless one is seeking the esoteric, anyone will do as one's neighbours. The beauty of the typical is that one finds it all over the place. But is it really so simple? For as anyone knows who has lived in a large and variegated village, as soon as one has settled in, one is associated with or adopted by a particular section of the community; as one becomes an insider to this section, one becomes an outsider to the rest – true, no longer an outlandish unclassifiable alien, but a member of a different faction or clan to be treated as different and told different things. One is not regarded as a neutral observer because such a category does not exist. So how can one maintain the pretence of neutrality? Our allotted place in social space largely determines our subsequent experiences and what we are privileged to observe.

This is not to say that in residing with a chief or a priest one assumes a chief's or a priest's perspective on society, for the ethnographer is more than a transcriber or mouthpiece for his or her informants. It is merely to say that we begin by sharing our hosts' or our friends' networks, and to recognise the implications of this obvious point. The way we are socialised into another culture, and hence our point of view on its internal variations, is bound to reflect our position in the social space of our host community, something partly beyond our control. Even if the two halves of that schizophrenic creature the participant-observer are analytically separable, in practice we are defined by a certain position which commits us to a certain way of acting, and this feeds back into the kind of things people let us know.

The epistemological problems of cross-cultural understanding are thus compounded by the practical constraints of fieldwork. To illustrate some of these points and the problems they pose for the writing of ethnography, I want to describe an event which took place in the early months of fieldwork in Nias in 1986. The event was a feast which I had planned as a kind of inauguration, but which turned out to be more of a rite of passage in another sense, allowing me to extricate myself from an uncomfortable association with one part of the village and – metaphorically and literally – to enter another. It was more than a change of position. The experience as lived, and as remembered or reconstructed later, were quite different. This has partly to do with the nature of the feast as a complex event; but it also suggests how what we know at a given moment is embedded in contingencies, how committed participant and free observer are yoked together, like it or not. The experience seems to me to be worth relating here because it is in certain respects characteristic. It belongs, roughly, to the initiation genre and its subgenre, the tale of entry. Like most such tales, it proclaims a double legitimation: the novice become anthropologist and the marginal figure incorporated and certified as honorary member of the host society. But its typicality is, as I shall show, qualified. It does not contain a revelation or sudden illumination as the form demands. We will see in fact what can be distilled out of this story and what remains enigmatic and unresolvable. I try to extract what ethnographic lessons there are, and examine what literary problems are involved in its presentation. But my underlying theme is sceptical. I suggest that some of our experiences in the field, perhaps those which seem to us most important, contain a residue which resists interpretation and analysis, and remain in the end obscure while seeming significant, formative without being informative.

A FEAST OF MERIT

The village where my wife and I lived for two years, from 1986–88, is in the hilly interior of Nias, an island off the west coast of Sumatra. It lies a day's walk from the trunk road which connects the small port towns at either end of the island. The last part of the journey is through the rocky shallows of the river Susua which enfolds the village in a broad loop. Unlike a Javanese village, which you can pass through almost without noticing, the Niasan settlement is, as often as not, isolated and difficult to approach. A sense of the layout is important for what follows. The upper hamlet, once heavily fortified, is built on a hill. Steeply roofed lineage houses fronted by rows of megaliths are grouped close together around a paved plaza. This enclosed space, inward looking but outwardly bristling against intruders, forms a kind of amphitheatre well-suited to the agonistic ethos of Nias life, a public space in which political action is not only carried out but witnessed and legitimised.

At the foot of the hill beside the river, a lower extension has sprung up in the last 20 years around a weekly market place, consisting of a half-dozen wooden or semi-concrete houses, a primary school, and a small one-man clinic. Christian since about 1920, the village shares a church – built in 1957 – on its border with a neighbouring village. Tradition and modernity are thus neatly (perhaps too neatly) contrasted and to some extent separated by the village layout. The teachers, nurse, priest and officials of the lower hamlet do not circulate in the upper hamlet; and the upper folk only descend for some specific purpose or on market day. The village headman, a traditional chief who had ruled the area for over 30 years and who died during our stay, was an anomaly in this respect. He had moved down to the lower hamlet a few years earlier in order to set up a small store near the market, but he was gradually withdrawing from active life.

As newcomers, it was natural that we should find it easier to associate with the more modern people in the lower hamlet. Language was no barrier as all had some education and spoke Indonesian; and it was easier to explain why we were there. On an earlier reconnaissance, accompanied by a guide, I had fallen through a log bridge a mile from the village and had been obliged to convalesce in the house of the hamlet head – Ama Yuco, as I shall call him. During my first week in the village, I was more or less confined by my injured leg to the lower hamlet and was constantly in the company of this influential man. I was grateful for the delay in plunging into fieldwork proper – whatever that may turn out to be – and for the opportunity to witness the many meetings and discussions that took place at his house, though without understanding anything save through his later explanations in Indonesian. I was given to understand by him and his visitors that he was the real authority in the village and that the ailing chief, who lived in a small house opposite, was now only a ceremonial figurehead. Nevertheless, against his advice, I accepted the offer of the chief's old house in the upper hamlet and moved in after fetching my wife a few weeks later. It was a traditional chief's house, a massive edifice built by the whole village, strategically and symbolically located at the heart of village life. Living in it put us under the patronage of the chief, but we were far enough away not to be entirely at his mercy. It also allowed easy access to both parts of the village and gave a broad view of the general scene.

However, despite this professional desire to be at the heart of things, in those early weeks we were forever scrambling down the slope to the lower hamlet. Among the upper denizens I could not get beyond pleasantries in Niasan, and – or so I reasoned – I simply could not delay 'data gathering' until I had a better grasp of the language. In the lower hamlet, my halting efforts in Niasan could be made good with Indonesian.

We settled in without ceremony. This in itself was slightly disgraceful. As people liked to tell us later, with a special knack Niasans have for elaborating on an offence, 'We should have feasted you, killed pigs, and announced you to the village.' We later learned to reply to this kind of challenge that it was indeed pretty bad, a sign of the times and a personal affront, but nothing so bad that a good portion of pork wouldn't put right. In fact, it was we who held the feast of welcome, planned as a kind of launch, some two months into the fieldwork. By then it was clear that personal standing in the village rested as much as anything on one's feasting record. If I wanted to be taken as more than a lightweight and if my enquiries were not to be brushed aside I would have to kill pigs. A proper *ovasa*, or feast of merit, demands at least 40 pigs; up to 100 or more for a big feast. But as a début three or four would be enough, a trickle that should later swell into a sanguinary flood. Any more now would raise problems about contributors, helpers and recipients, since all these roles are either duties or privileges associated with kinship and alliance; and one purpose of the feast was precisely to assign us positions in the structure: me, as member of the chief's lineage, my wife Mercedes, as member of the other major clan in the village. My nominal wife-givers took this idea quite seriously, expecting an avalanche of pork and gold throughout our stay. In the Nias system of alliance, wife-givers are a group whereas a wife-taker's debts fall to the individual alone, not to his lineage; he can, in turn, however, take from his own group's wife-takers. Since I had no close wife-takers, that is, no daughters' husbands or sisters' husbands within 10,000 miles, I could not easily recoup my customary debts by squeezing them in turn, and I was loath to tap into the chief's vast and lucrative network as I was entitled. It was a situation rich in possibilities for genealogical disputation. I spent much time quibbling over kinship terms, fending off solicitory chickens and blocking demands for pigs. This is something one should be able to do anyway in Nias.

The feast was suggested and organised by my initial host, the hamlet head, a cousin and lifelong rival of the chief. I let the feast happen and provided the means and some of the labour. Since there was no traditional basis for it, no one much wanted to participate except those who could expect something out of it; and it was too small to justify any of the usual performances, war-dances and the like, which would attract wider interest. In intention it was vague but well-meaning; in execution it was a shambles.

Feasting in Nias consolidates influence and wins prestige; at the very least it gives one the right to speak and be heard (see Beatty 1992). This is because, as a general rule in Nias politics, no one will listen to you unless you can match deeds to words. If fieldwork is an initiation into academic discourse, so, in Nias, feasting is an entry into public debate. It was one of my neighbours who pointed out this analogy to me, adding: 'If you haven't followed the path of suffering will anyone listen to you?' Feasting, like fieldwork, is not simply a purchase of rank or a mere ritual attainment: it is a kind of proof. Thus, in order to carry off a major feast successfully one needs more than wealth and good intentions. A full-scale *ovasa* is a feat of organisation, string-pulling, manipulation, oratory, prevarication and political nous. In this sense, as I learned to my cost, overall control cannot be delegated. The feastgiver needs a general's sense of what is going on in different parts of the battlefield. As he walks among scores of carcasses surveying the butchering and weighing of prestige portions, he must see to it that no one is left out, that the right people get their due and in the right order, and that little is filched. In short, he must see to it that, although no one will admit it, everyone is happy. In the fine details he must be able to trust his lieutenants. But he also needs to maintain a clear view of overall strategy: behind, or rather through, the exchange of prestations – the conferring on him of affinal blessing and noble title, the influx of tribute from wife-takers and other guests, and the circulation of prestige-payments – he is busy paying off debts, incurring others, punishing ingratitude or neglect, rewarding and obtaining loyalty, redrawing the political map.

But the *ovasa* is not self-contained. Each *ovasa* relates to or responds to previous ones, not only in the pattern of debts, but in theme and political import. And the principal guests, the affines and the rival leaders, have their own agendas: their desire to reward or slight, to eclipse rivals or win allegiance among their supporters. Each major gift of tribute triggers or climaxes a sequence back home of which the host may know or care little. The exchanges focused on the present *ovasa* radiate out in time and space to numberless other exchanges of varying significance to the present event. An ethnography, if it is to reflect this complexity, cannot limit itself to a single event or a single perspective, be it that of the host or even of the observer. Of necessity it is a reconstruction of various perspectives of different participants, and of various related events. There is no such thing as 'the *ovasa*', only actual *ovasas*. But actual *ovasas* cannot be grasped in their entirety since they are not self-contained; they ramify outward throughout the whole society. The particular features of any *ovasa* make sense in relation to these ramifications.

In miniature, these considerations – of organisation, trust, strategy and so on – applied to our inaugural feast, magnified by our inexperience and the novelty of the event for everyone, which meant the guidelines were extremely flexible and were bound to be flexed according to personal advantage. After only two months we had little sense of how pervasive and how morally and politically charged were the debts among villagers, and therefore how unlikely it was that we could hold a neatly bounded feast in the way we wanted merely by buying four pigs and two sacks of rice, and declaring our intentions.

Remember, we had moved into the upper plaza, but still had at least one foot down below in the lower hamlet. The hamlet head, Ama Yuco, since our settling in, had taken a close, almost obsessive interest in our welfare. Like others, he would advise us about whom we could trust, who would answer our enquiries truthfully, whose betel or coffee we should refuse in case it contained poison. (A small triumph later on was the day I drank coffee in the house of the official Village Poisoner.) Although I was no longer the scholar-parasite of the early days, seeing the world through his eyes and dependent on his translations, somehow Ama Yuco always knew where I had been, who I had talked to during the day, and often, in surprising detail, what had been said. But the relation was changing. We sometimes ate at his house; my wife was exchanging language lessons with his grown-up daughter; I had helped him construct a fishpond and had bought the underlay in town. We were friends. Nevertheless, we sometimes felt uncomfortable, even alarmed, at his constant interest. Sometimes I would arrive home after a long afternoon of doing the rounds and find him waiting for me, chewing betel and staring out of the window, eager to sift through and comment on my findings. Or he would sit for hours next door in the house of a much-despised man he could impose on in this way, waiting for my arrival. He rarely ventured further up the square, preferring to let people come to him; but he seemed to have a hand in everyone's business and knew everything that was going on. It was said that he had many debtors.

Like all the half-dozen leading men in the village, he disdained attendance at church, regarding it as a platform for social climbers and a place of hypocrisy. But unlike anyone else I had met he could talk about Christian ethics in reassuringly familiar terms. The Old Testament prohibitions and fear-mongering espoused by the Nias Church were, he said, a distraction from the real dilemmas posed by Christianity.

He loved to expound, usually to an audience of upturned gaping faces, the positive virtues of compassion, common humanity and charity so lacking in the darkness and atavism of Nias. There was a certain fascination in these monologues; as he strode back and forth grimacing and gesticulating before his audience, one had the impression of a man wrestling with his demons, trying unsuccessfully to persuade himself of an argument that was contrary to reason. The point about Christianity, he would say, was that it required one to practise the impossible. In contrast, there was a chilling conviction in his interludes of cynicism, whether he was dissecting the motives behind an apparently kind action or turning over some theological point. 'Perhaps', as he once confided to me in an unguarded moment, 'the Bible is all made up anyway. After all, we here in Nias, are at best God's stepchildren, disinherited and condemned to poverty. Why should we believe the promises of men of another time?' This scepticism, expressed like my own with caution in the dogmatic rather than pious ambience of modern Indonesia, was a small area of common ground. But I rather think it was his unusual ability to view his world askance as one of many possible worlds, and hence his sense of our predicament as outsiders, which had led him – informant and mentor – to select us.

Ama Yuco had a pretty dim view of his fellow villagers. In his view, they were ignorant, too stupid to be wicked, but greedy and dishonest, out for what they could get. On the eve of the feast, as we were preparing the food, he took me to one side and warned me that two of our neighbours – one of whom I had crossed but the other, I thought, an ally – had approached the religious teacher with plans to get us out of the village. They had said there was no profit for anyone in our remaining there. The religious teacher had let Ama Yuco know, as an old crony; moreover, he had a hand in the present feast as supplier of one of the pigs so he would defer taking sides. I never got to the bottom of the story, but later suspected that Ama Yuco had seized upon it as a way of isolating us. At this particular moment it was unsettling, to say the least, and there was little we could do. The key person in any decision to oust us would be the chief. We had neglected him but it was now too late to involve him somehow in the feast. He was to be the most honoured guest and so it seemed hardly appropriate to bother him with the details, but he was a kinsman of the disaffected neighbours. The feast was unravelling before it had begun.

Nevertheless, preparations went ahead: the fetching of water, the pounding of rice, the rolling of betel, the warming and folding of banana leaves for wrappers. Ama Yuco's efforts to recruit help failed, to my surprise. Everyone's children and wives were, apparently, sick or unavailable. So it was up to us and his family to do it all. 'Never mind', he shrugged. 'We won't have to give away any pork in payments.' He had decided, or rather persuaded me to decide, that the preparations were best done at his house. It was nearer the river and and he had a ready supply of firewood. There would also be more control over who came and went, fewer hands on the pork, and less gossip about who got what. It seemed to me that he knew what was best or what he wanted, but I had to discern this and make each decision myself. In the event, the preparations were almost secret. Only when the pigs were tethered to be killed did other people appear. (Butchering is an expert business and must be witnessed.) The religious teacher's pig turned out to be smaller than promised and paid for – a subject commented on in the aftermath of the feast. When the actual carving up and division began, it was so minutely detailed, prescribed down to the quartering of the jawbone and the lengthwise splitting of the tail, that it seemed impossible anything could go astray. The rice, too, had been calculated to feed over 100. Since the feast was to be held in the upper hamlet, the individual portions had to be wrapped instead of served directly on banana leaves as is customary. In the late evening we carried the parcels in baskets up to the chief's house and pounded the huge gongs which swung from the rafters.

Very slowly the vast building filled up. A few youths strolled in out of curiosity. I could see men smoking outside, sitting on the stone plaza, taking no interest. The ubiquitous religious teacher and his choir struck up a song to the half-empty hall. Ama Yuco looked ill. I felt sorry for him and guilty at his predicament. Instead of deriving satisfaction and some prestige from having helped us, everything had turned out badly. He said that he was sickened by the lack of solidarity in the village. But evidently few people knew what was happening. Villagers would come as soon as they could smell pork, he said scornfully, but when they didn't he had to go after them himself, from house to house round the square.

Almost last to arrive, accompanied by his wife, was the chief, haughty and punctilious, dressed as first district head in a uniform given him by the Dutch. Seated on a podium he scowled down at the villagers, muttering irritably at children and issuing minor commands. His barbed jokes reminded us that though he may be ailing he was still the chief. I had prepared and memorised a simple speech in Niasan. The village secretary, understanding more or less what I wanted to say, had rewritten it idiomatically to have the desired effect and had made some interesting additions. Nevertheless, it was a dull little speech, so I shall quote only the curious concluding words:

We chose this village because it is famous in Nias for its good custom, and its great and knowledgeable men. We want others to know of it in our country so they can learn your ways. There is no difference between how we do things and how you do things. It's exactly the same where we come from. Although the language is different, there is no resentment between us, truly no resentment. So be patient with us. There's no food here today, only hot water, one glass only and nothing to put in it. We beg your indulgence if our faults are great. Ya'ahovu.

There were nods of approval and some head-wagging surprise that I should know about such things. Our superficial differences did not inevitably mean resentment and envy. Underneath we were the same. As one elder said to me, 'Under heaven we are all God's pigs, though you're a white one and I'm a black one.' The chief, pleased with my praises of his greatness, then rose to his feet and began a full-blown oration, berating the villagers for their lack of courtesy to us, echoing my praises in shameless boasts of his achievements, and generally softening up his audience with a mixture of jokes, reproaches and good-humoured banter. Each phrase was punctuated with a long thrilling call from his respondant on the opposite side of the hall, a kind of antiphony to the main theme. The speech culminated excitedly in the whole assembly chanting our new clan-names followed by a ritual shout. When the chief sat down next to me, there was a gleam of satisfaction in his eye at a job well done. The old lion could still roar.

The evening immediately took on a livelier, almost party, turn with singing and dancing. The headmaster, addressing me as 'brother-in-law', sang 'Have I told you lately that I love you'; an elderly man sang the Japanese anthem, learned during the occupation. The religious teacher put on his spectacles and in a strident nasal tone performed a psalm. While all this was going on, I went to the rear section of the house and found a group of men squatting in a dimly lit circle around piles of meat laid out on palm leaves. There was much fussing and rearranging of the pieces. So intent were they on the division that it

was some time before one of them looked up and noticed me with
what I thought was some embarrassment. Not for the first time, I had
a curious sense that it was not my feast; I was an unwelcome onlooker
and went back to the clamour of the big hall. The food was brought
out; the little wrapped parcels of rice each now containing a piece of
pork whose size and recipient had been determined by the cabal at
the rear. Not many guests had been served when Ama Yuco prompted
me to give the signal to eat. 'Manga ita', I said to the bewildered crowd.
We unwrapped our packages and were dismayed at the tiny portions
of pork, praying they were not typical. As more parcels came out there
was a general motion towards the door which became a scramble as
if the guests had been dismissed or dreadfully insulted. 'Don't worry,'
said the chief kindly, 'they can't stand a long session and they want
to take their food home to share.' Ama Yuco, by now sullen and ashen-
faced, pronounced a customary formula of thanks to the departing backs:
'There is no return from us.' Then he too made his excuses and left.

For the next few days, oddly enough, things continued as normal.
No one mentioned the feast to us. The whole thing had died quietly.
But I was still puzzled and annoyed that certain important guests had
not attended and resolved to find out the reason. My first call was the
man who had been teaching me *sile*, the Nias martial art. Ama Yeri
lived in a house set apart from the rest in the lower hamlet. He was
a prickly character, a bit of a loner and a staunch opponent of the church.
His house was an occasional gathering place for malcontents, and the
views that emanated from it, focused through his few carefully chosen
words, had a definitely unorthodox slant. Somewhat pompously he
informed me that he had stayed away from the feast and had earlier
declined to help because he had not been formally invited. (My
personal invitation evidently didn't count.) However, seeing my
irritation at his excuses, he signalled to the lodger to shut the door
and leaned forward speaking in a whisper. He had learned, though he
would not say how, that about half the rice had gone missing. The
religious teacher had made off with a good deal of meat and had even
paid off creditors. Others whom he wouldn't name had benefited too.
As a result, the village elders had not been properly rewarded; some
of them were insulted. He tore up pieces of betel leaf and laid them
in rows on the floor, indicating the elders and their respective portions.
The slaughtering and butchering should have been done under the
chief's supervision: he would be offended to have been left out. After

all, the feast was in his house, and we were his official guests. The butchers and the pig-sellers should not have claimed any extra portions for their services. The division of cooked meat should have been done in the front public hall in front of everyone. And so on. Above all, I should not have trusted Ama Yuco. 'Don't trust anyone except the chief', he warned. So what should I do now? Should I confront Ama Yuco? 'Don't show any sign! Be the same as usual. Keep quiet about what I've told you.'

That same evening two men of the upper hamlet came to see us, on the pretext of borrowing betel leaves. 'The village is angry for you', said one of them. 'Why didn't you ask for our help instead of leaving it to those two rogues? It should have been done like this.' And he repeated the demonstration with the betel leaves. 'Fancy trusting the religious teacher! He even steals the choir's portions.' Again we were warned to behave as if nothing had happened and not to risk a confrontation.

I spent the next few days going round the upper hamlet testing opinion and mending fences. As the story was told over and over and particular details were highlighted and embroidered it was clear that the blame was shifting from the organiser, Ama Yuco, to the religious teacher, who had in fact played only a secondary role in the event. There was an evident glee in relating how he had cheated me over his pig, recalling other deceptions and broken promises; how, with poetic justice, he had then bungled the killing, having to stab the terrified animal repeatedly; how his choir could only be persuaded to attend with the promise of pork and had performed miserably.

Yet it was only those with an acknowledged grudge or long-standing enmity with Ama Yuco, the prime mover in the feast, who would directly blame him. Unlike the religious teacher who had married into the village, he was a native, a powerful patron at the centre of a web of debts. His eloquence in public debate was admired but much feared by opponents. Behind this practical worry there was a reluctance to criticise a leader, the cousin and likely successor of the chief. Chiefs are not expected to be saints, and no one is surprised by their venality; but leadership has a sanctity which cannot be questioned. 'Why didn't you warn me what was happening?', I asked a lineage mate who had watched the butchering. 'We were afraid', he said, pointing upwards. (Was it with irony? I could not tell.) 'Afraid of God.'

The role allotted to us in these retellings became that of dupe. It was an inglorious role, very far from what was intended; but it marked a kind of induction none the less, since many others had been duped by the same men. We were in good company. As we struggled to respond to the practical consequences of the feast – of losing a key informant and of readjusting to the world of the upper hamlet – it was the psychology of the protagonists that obsessed us. Why had they done it and risked their reputations? I think it possible that both men, profoundly moral in different ways, persuaded themselves at the outset that they were doing us a favour and reaping a moderate, if illicit, reward, at the same time as paying off a few old scores by not inviting their rivals and enemies. On the other hand, they might think, as cynics suggested, that we wouldn't appreciate the harm, and that those who did understand wouldn't care: as outsiders we were fair game. Never mind that the feast was intended to make insiders of us. On reflection, though, it seemed that Ama Yuco more than anyone would feel keenly the dilemma posed by this irony. What Christianity represented to him, as he had often explained to me, was a near-impossible ideal of a universal morality, an ethic in which there were no outsiders. As he put it, we were equals before God and our salvation depended upon our acting towards each other as equals. Yet we were not equally endowed by God; we faced each other across an invisible divide. 'Here are you,' he said, 'able to fly half way round the world to satisfy your curiosity, while my children will never see beyond their Nias village.' God's stepchildren indeed. His own cultural wealth, newly revealed by my interest, contrasted painfully with his material poverty. What could it mean, then, to help me, an outsider, an equal of sorts, join the village? What would be his profit or loss, financial, social and spiritual, if he frustrated that aim to his own advantage? His expressions of growing disillusionment and disgust as support had melted away now struck me as ambivalence, even regret, about the course of events. One particular moment stands out. It was just before the guests started arriving – too late to undo what had already been set in motion but before any public offence to myself or the guests had been given. We stood in the vast empty hall of the chief's house. Ama Yuco, somewhat wearily, asked me when I wanted to collect the legs of pork due to me as feastgiver; they were still down at his house. I said we wanted him to have them as a token of gratitude. Ama Yuco appeared puzzled and unusually lost for words. His lean sallow face, normally animated with the practised expressions of the orator, was lifeless and grim. 'It would be like payment', he protested, but with a note of hesitation which invited me to rephrase the offer. 'We want you to accept it as tribute to a senior kinsman, since you cannot be among the guests yourself.' He looked at me sternly for a moment and then said, 'Ah, well, in that case', and shuffled away awkwardly. At the time, I wondered whether I had compromised him by denying him the opportunity to play the part of patron and helper: it was a reward which, as a poor man, he could not afford to refuse. Later I saw the offer as a turn of the screw, cruelly if unwittingly applied, replete with an irony I could not have known.

What can one learn from such an experience – one which is in some ways so typical of fieldwork accounts? Does it illuminate anything beyond the predicament of the fieldworker? Someone determined to salvage something for a general account of feasting would probably treat it as a negative case. Let me briefly do so, if only to emphasise what would get left out and to underline my sceptical point about the enigmatic nature of experience in the field.

The essence of a feast, even a small one, is as a display of generosity. Only when the floor is carpeted with rice and pork does the host stand up and with mock modesty apologise for the humble offerings, inviting the salivating guests to eat the 'meagre sweet potatoes' piled before them. A good Niasan host knows that enough is never enough: one has to overwhelm. But, ideally, wealth is hidden until the opportune moment when it can be ritually disposed of to the maximum credit of the host. We had reversed this procedure, squandering in the early stages without ultimate credit. Perhaps a third of the food had been consumed or salted away before the event even began. The feast itself was marked by concealment rather than display, reversing public and private space. The wrapping of food parcels had, unknown to us, disguised the smallness of the portions; the grading of portions according to status had been done at the back, out of sight, instead of in full view; finally, to compound the reversal, the food was taken home to be consumed. The point of a public division is that (a) it is difficult to filch anything, (b) status is displayed and acknowledged and (c) there is an audience of knowledgeable men. This makes it less likely that someone will get overlooked (except as a calculated insult). Moreover, in the normal course of a feast, the protracted fussing and rearranging of heaps of meat has a useful function beyond making sure that there is enough to go round: it forms a kind of preliminary negotiation of who gets what. The meat dividers, necessarily expert and thick-skinned, are subject to a stream of unsolicited advice from the heads craning over their work. Amid the thud of chopping machetes, claims and counterclaims fill the air. Someone reminds a butcher that he is representing three brothers; someone else that he has come from afar. Beneath this hum of informed but biased comment, the butchers, squatting around the meat and taking no obvious notice, quietly ponder the portions, adding an ounce here, deducting a bit there, swapping a shoulder cut for a piece of belly, reassigning an elder's portion to a junior. If all has gone well, when the prominent men carry off their trophies and the rest collect what remains, there is, to use the Nias phrase, no resentment.

The problem of how to divide the meat was compounded by the novelty of the feast. I had no debt record; I did not fit into a pattern of exchange, so there were no clear guidelines, only a barrage of conflicting advice. My disaffected neighbour, a junior member of the chief's lineage, suggested that the portions be made equal and to do otherwise would show favouritism. But that would mean slighting the men of high rank whose feasting record entitled them to a bigger share. Then there was the problem of whether my nominal wife-givers and lineage mates would assert their claims over other villagers; and would my non-kin countenance these claims? Clearly there was no neutral position and no neutral advice to be had. I left it to the butchers to determine, trusting in their judgement and hoping thereby to avoid the blame. I was later blamed precisely for giving them a free hand.

Evidently there are certain lessons to be learned from a negative instance such as this one; though, of course, one only recognises it as negative retrospectively in contrast to other cases. But the lessons are relatively few, and, on the other hand, my personal involvement made the event unique and unrepresentative of Nias feasting. Contrary to Rosaldo's credo, experience was a barrier as much as a clue to under-standing. Other feasts I witnessed or took part in as a minor player displayed a much clearer pattern and predictability of outcome; they had a clearer focus, without being rigid or completely standardised in form. If there was a deeper lesson, it was to give due consideration to the pragmatic aspect of feasting – not only the strategems of exchange but the peculiar moral intensity which informs particular transactions and which totalising structuralist accounts of exchange generally neglect. However, what made the feast formative for me was not its power of illuminating other events. It was the intensity of the encounter, the sense of make-or-break and the impulsion to reorient. The feast was not merely a fieldwork experience but an experience about fieldwork: of finding the right words and the right voice, of learning how to be neither oneself nor what one's hosts would make of one but something in between. Above all, it was emblematic of the reflexive confrontation with the informant as alter ego – everpresent shadow, *Doppelgänger*, a friend, a conscience, finally (and mutually) a burden.

The ethnographic rewards, then, were small. But the event remains for me challenging and enigmatic for other reasons. Not because some crucial bit of anthropological data was missing, but because of the complex psychology of the drama, the ambiguous and discomforting politics of the situation, and the sheer difficulty of establishing who did what and why. These are essentially literary, narrative problems, a question of reconciling different versions of the 'same' thing. To some extent they must adhere to any such complex event; but they loom large in those events in which we ourselves are central actors.

Insignificant as the feast was in local terms, there were several versions of events in circulation. Ama Yuco, no doubt, had his own view of things, as did the religious teacher. (Since we have separate audiences, we can each claim the last word.) But there were other, distinctive, opinions coming from different positions: the disgruntled non-attenders, like my *sile* teacher, whose account was based on hearsay – and on what he wanted me to think or thought I could grasp; the chief, a splendid egoist who saw in the event no offence other than to himself; the men of the upper hamlet – juniors, annoyed at favouritism, seniors, insufficiently acknowledged. There was the apportioning of blame, shifting according to political expedient and audience. Somewhere in all this was my own view or views as, in varying amounts, participant and observer. And of course, events took place against a background of continuous comment with my wife, who shared the experience and endeavoured with me to make sense of it: the first unwritten stage in ethnographic translation.

RETROSPECT

How should one represent all this in a coherent account? Does a coherent account belie the muddled and improvisatory nature of what is happening and the plurality of competing perspectives? The literary device of the unreliable narrator may be a solution to the novelist's problem of representing social reality, but it can hardly be a way out for the ethnographer. Moreover, the partiality but relative clarity of a unique point of view, that of the observer, is further muddied by doubts about the nature of participation. Whose feast was it anyway? Proposed and organised by others for their own reasons, interpreted and represented in a variety of ways – even my speech was written for me in a language I could not yet control and in words seemingly at odds with my purpose. If it were true that there was no difference between how we do things and how they do things, what was the point of being there? At the very moment of 'entry', as I spoke these curious words, I was most conscious of the cultural difference they denied. And if my own motives and actions were deeply ambiguous, what was I to make of the part played by others? Clearly there can be no end to such speculation. What, then, do these doubts imply about the validity of ethnographic observation? If such doubts as these cloud judgement in the case one knows best – oneself – then surely they must apply even more severely to the less accessible experiences of others. Yes, if it is the experiences of others which we are after. But it will be clear by now that the peculiar difficulties I encountered are not really typical of anything except fieldwork. They are implied in the paradox of participant observation and the difficulty of, as it were, trying to see round one's own corner.

Let me finish, then, by summing up my position. I want to question the assumption that the trajectory of fieldwork experience and the growth of ethnographic understanding run in parallel. The one is marked by dramatic ups and downs, tied to the present; the other is gradual and recognisable or partially comprehensible retrospectively. How we come to learn another culture is a far from transparent process and it cannot be tied to the vicissitudes of experience. This gradualist position entails a rejection of a reflexivist privileging of dialogue. I want, further, to question the assumption that by writing ourselves into the ethnography we are somehow making everything plain, being more honest; that we are showing the reader not only the facts but how, in dialogue, the facts are constructed (Rabinow 1977). Such an approach, defined in opposition to a naive empiricism, retains something of its optimistic illusion of experimental control. Instead, I argue that some of our most critical experiences in the field remain opaque and resist analysis. Like a vivid dream they impress us as significant; they demand interpretation but there is not always a ready method at hand. Unlike the dream, meaning remains elusive not because it is buried in the unconscious (though, no doubt, this plays its part) but because it is inaccessible for other reasons bound up with the paradox of participant-observation. The moment is past and perhaps survives only in the accounts, more or less garbled, of others. The meaning of the event may pertain to others' motives which are beyond investigation; it may require a linguistic competence we do not yet possess; it may be more complex than our provisional interpretive models can comprehend. And of course, we can hardly claim full knowledge of the historical contingencies, let alone the immediate ones, which determine our situation. As protagonists in the drama we, like our hosts, can fall victim to a dramatic irony which later, as authors, we attempt to finesse. Significant but obscure, such crises become for us personal symbols, and it is their evocative power as symbols, enhanced in the telling and retelling, which gives us the illusion of an epiphany. We feel something very important has happened; indeed it probably has; but what to make of it? Such formative experiences mark us as anthropologists and influence the way we look at fresh problems. But we assume too much self-knowledge (and other-knowledge) if we ground our ethnographies in them. Tales brought home from the field are the tribal valuables of the profession, crafted and polished until they have the patina of authenticity, handled until they are redolent with personal *mana*, oft-exchanged but never fully alienated from their owners. Let us respect their mystery without needing to revere them.

NOTE

1. Geertz writes:

   It was the turning point so far as our relationship with the community was concerned, and we were quite literally 'in' ... It led to a sudden and unusually complete acceptance into a society extremely difficult for outsiders to penetrate. It gave me the kind of immediate, inside-view grasp of an aspect of 'peasant-mentality' that anthropologists not fortunate enough to flee headlong with their subjects from armed authorities normally do not get. (Geertz 1973:416)

   A later essay rejects this equation of acceptance with illumination. Where the early formulation aims at the 'inner nature' (whatever that is) of a society (1973:417), the later one, more cautiously, speaks of 'natives' inner lives' (1983:70).

REFERENCES

Beatty, Andrew (1992) *Society and Exchange in Nias*, Oxford: Clarendon Press.

Geertz, Clifford (1973) 'Deep Play: Notes on the Balinese Cockfight', in *The Interpretation of Cultures*, New York: Basic Books.

Geertz, Clifford (1983) '"From the Native's Point of View": On the Nature of Anthropological Understanding', in *Local Knowledge*, New York: Basic Books.

Rabinow, Paul (1977) *Reflections on Fieldwork in Morocco*, Berkeley, CA: University of California Press.

Rosaldo, Renato (1993) 'Grief and a Headhunter's Rage: on the Cultural Force of Emotions', in *Culture and Truth*, London: Routledge.

4 LEARNING TO BE FRIENDS: PARTICIPANT OBSERVATION AMONGST ENGLISH SCHOOL-CHILDREN (THE MIDLANDS, ENGLAND)[1]

Allison James

INTRODUCTION

Hilaire Belloc, the master of cautionary tales for children might well have paid heed to the tales children tell amongst themselves. Adult warnings about the foolhardiness of eating string and telling lies pale into insignificance when set side by side with the grim stories children tell one another of the consequences which can follow from particular strategies of action. Pondering the transition to middle school at nine years old, Christine offers one such cautionary tale:

When my sister Jane went to middle school these people ... her friends ... loads of people said and when they put their chairs up [on the desks] and it fell off and bumped [Susie's] head and Susie said [that] it was all blamed on my sister. And Susie didn't see [it happen]. And it was Susie's friends just making it up to make Jane not liked. To make Susie not like Jane and when they were going [home] Susie's friend stamped on her foot and Jane got told off again, And when the taxi came she ran in crying and I wondered what was the matter with her and she told me all about it. And my dad's had a word with her teacher.

By no means a central figure in the group of girls who are chatting with me, Christine is none the less moved to speak. Hesitant and inarticulate at first, her tale becomes more fluent as she warms to her task of depicting the intricacies of girls' friendships with one another. In the relative safety of her familiar primary school, she offers a caution for the other girls, a warning of what might happen to them when,

98

no longer juniors, they enter the tougher social world of a middle school. At the same time, however, her tale confirms what she and they have already got to know about the complex process of being, becoming and losing friends.

How I came to learn about the process of making friends in childhood can also be seen as a cautionary tale of sorts. Through its telling can be glimpsed a methodological critique, alongside an account which, cautioning against singular explanations of the social world, reveals the interpretive value of a more polyphonic style achieved through reflecting not only on my field notes but on the process of 'noting' the field (Rapport 1991). This confirmed for me the value of the ethnographic method when working with children and affirms the insights generated by using 'thick description' to interpret their social world (Geertz 1975). The tale I shall tell reveals how, as I watched and listened to the children interacting with one another and with me, they literally enacted for me (and for each other) some of the structuring processes through which the culture of childhood is given form and meaning. For all of us, therefore, the incidents I describe here were educative, providing a lesson in 'practical mastery' for the children and one in 'symbolic mastery' for me (Bourdieu 1977).

A PROLOGUE

To recount this drama, I shall use an appropriately theatrical metaphor and begin with a prologue which tells an everyday story of childhood. But it is not a tale of beguiling innocence and charm. Such myths about the nature of childhood, which as a culture we have spun for and on behalf of children for so many years, I had already long abandoned when, in 1993, the murder of one child by another threw contemporary conceptions of childhood into such disarray (James and Jenks 1996). My first fieldwork in the late 1970s in the north-east of England, together with a further stretch amongst four- to nine-year olds some ten years later in the Midlands, had taught me to see that childhood, for many children, is not the safe, innocent and protected childhood which adults fondly mythologise (Holt 1975; Ennew 1986). Both experiences have made me conscious of childhood's darker aspects; buried memories fleetingly and painfully resurfaced and I was forced to confront the social construction of childhood in the often anguished

moments of its making. That is to say, fieldwork made me relearn what
it is to be a child through witnessing children learning how to be children
in the same moment as they were learning to be themselves. As they
negotiated, manipulated, kicked against and submitted to the social,
economic and political limitations placed upon their actions by the
adult world I saw how, as individuals, they came to some understanding
of the meaning of the category of 'child' and of the social time and
space deemed 'childhood' within which their notions of selfhood were
temporally and temporarily confined. That is to say, I was led to
understand that children must learn forms and styles of behaviour
appropriate for being a 'child' and for themselves, as particular children
in particular settings. In brief, therefore, my work has become the
recounting of childhood as children experience it.

However, this understanding of what it means to be a child has been
– perhaps fittingly – a long time maturing. Beginning in the late
1970s, my initial interest in children developed through an enquiry
into the representations of childhood to be found in children's fiction.
This later translated into a desire to explore, through fieldwork,
children's own account of what it is to be a child. In contrast to the
passivity imputed to children through traditional socialisation research,
I followed Hardman's (1973) lead in seeing children as competent social
actors who could be articulate about what the social world is like for
them. It was this approach which became a central feature of the contemporary 'ethnography of childhood' movement, which first began
to coalesce in the early 1980s (see James and Prout 1990), and which
underscored the socially constructed nature of childhood. Although
acknowledging its biological base it is argued that childhood is primarily
a social institution which contextualises the process and outcomes of
children's lives and one which therefore varies between and within
cultures. It is these social experiences which ethnographies of childhood
seek to capture and, in a more mature contemporary 1990s account,
is acknowledged the fact that to write about the social construction
of childhood from 'the child's' perspective, is not to make claims to
reveal the authentic child. More humbly, it is to provide a rendering
of what childhood might be like (James 1993).

This, then, is my engagement with childhood: in sum, I reject the
inherent biologism of accounts of childhood that see children as the
passive recipients of culture into the next generation and offer, instead,
a view of children as active participants in the social world. This is a

view of children as seekers after social knowledge and makers of social relations, of childhood as the social and temporal context of children's daily lives which shapes who they are and will be, in the same moment that, as children, they help construct what form the institution of childhood will take in any particular culture.

However, my present understanding of children's experiences of the social world has been reached but haltingly, by a series of by-ways and back-waters between which I have freely zig-zagged (Rapport 1993). This chapter traces out part of that ongoing intellectual journey. I chart here, what were for me, a series of salutary lessons during my last period of fieldwork wherein the dense web of children's social relationships was literally teased out for me by the children. It was an experience which ricocheted through my fieldwork memories, providing a point of critical reflection both during and after fieldwork in the manner described by Okely:

Where the anthropologist continues to insert (or reflect upon) the particularities of her discussions through the length of the field experiences, the material does more than describe the type of relations between the anthropologist and the people concerned. We are also able to see how the interrogator acts as a catalyst in eliciting defining aspects for specific members. (1992:14)

In what follows, a number of issues about the process of interpretation are raised as I consider particular events in the particularities of their social and emotional contexts and then reconsider them at a distance removed. As interpreters of the social world for and on behalf of others how do we arrive at our interpretations? How can we account for the conceptual distance that often arises between initial hunches, inspired or tempered by emotion, and later more considered and detached reflection? How might bridges be built between these interpretive acts, acts shaped by different moods, a different time or a different intellectual environment? And what do the gaps and lacunae between these different interpretations reveal about the nature of our subject and the process of anthropological explanation? This chapter addresses some of these questions.

THE SCENE

That my own anthropological training was inspirational rather than methodical perhaps accounts for my curiosity about methodology.

Lacking the experiential rigour of following a methods course, it was through the tremendously personal and intellectually supportive relationship I had with my supervisor that I learnt to do anthropology. It was with his guidance that I began to focus so intently not just on what children said but also on how they said it and it was he who taught me early on the importance of observing the minutiae of the mundane. Thus my anthropological work with children has been increasingly characterised by a concern for the fine details of children's everyday lives which shape the ebb and flow of their social relationships and which provide clues about 'what [children] are up to, or think they are up to' (Geertz 1975:15). Such a research agenda fits easily into the unremarkable daily routine of an English primary school.

Thus, July 1988 saw the start of my second piece of fieldwork with children in a large lower school in the Midlands. Interested in exploring children's ideas of disability and difference, I was keen to discover what children find different about other children and to what extent and in what ways differences are awarded significance by them in terms of friendship formation and daily interactions. I sought answers to questions such as: what do children notice as different about other children? What features about another's body or behaviour have the potential to become significant differentiating factors in one child's relationship with another? What makes such significant differences significant for only some, but not all children? Thus, as initially conceived, my research had an applied aspect, framed by a concern for how children with minor disabilities might fare at school.

But, as time progressed, the project altered course a little and I became interested not so much in identifying specific differences, but in the ways in which children use the idea of difference to mediate their social relations. Thus, what had started out as a practical project within the orbit of medical anthropology became framed by a wider theoretical exploration of the process of becoming social. In retrospect, I can see now that, in essence, I have been asking the same questions since 1977. But in 1988 I had yet to make explicit that connection.

The school I was working in had about 400 pupils aged between four and nine and was located in a large town in the Midlands. Its catchment area was quite diverse, excepting that it had a relatively low ethnic mix. The school had its own nursery unit and, during the year I worked at the school, I spent time with each year group. I had suspected that age and gender would be key variables in the attribution of identities between children and wished to capture the processes by which children gradually learn from their peers the special rules of naming and the allocation of discriminating distinctions which is so characteristic of children's culture. Moving from the nursery unit to the top class would allow me to document changes in the patterning of social relationships over time, relationships which I felt were shaped not only by children's growing physical and social maturity but by the experience of schooling itself.

However, soon after I began my research, which was being conducted through participant observation, to be supplemented with interviews, I became aware that the teaching staff had some concerns about my presence at the school. Clearly they were pleased to have me – I helped hear children read, dished out paints, cleared tables and could tell a story – but was I actually doing anything else at all? My unobtrusive observational style involved careful listening to the conversations the children had amongst themselves and with me so that I might remain alert to any subtle nuances of their language use. It required close attention to a body language which tells volubly of fallings out or warmer intimacies. Thus, I heeded their telling tales, laughed at their jokes, heard their complaints and rarely chastised or admonished as I helped with the papier maché and marshalled the cloakroom during PE lessons. The children regarded me variously as some kind of helper, as someone's mother, a sort of adult, a book-writer and sometimes as a friend. The different roles they assigned to me reflected my engagement with them at particular times and in particular places. As I got to know the older girls so they began to confide in me and to enjoy the chance to talk with an adult eager to listen and slow to condemn. The older boys paraded their skills for me to admire and although the younger ones would talk to me in the classroom, I would be temporarily estranged from them in the playground. So my relationship with different groups of children shifted in response to their concerns and affections.

The teaching staff, however, found my style of research unfamiliarly disquieting. Although unobtrusive and having little impact on their work routines, it lacked the outward trappings of more quantitative methodologies. Confiding in me their relief that I was not the kind of researcher who played on the swings with the children at playtime and sent them wild, none the less I could sense their unease. What was she about, this exotic anthropologist? And what was anthro-

pology anyway? Quickly, I set about devising tasks for the children
to do which could act as visible indicators of 'work in progress'. I little
realised how instructive these would turn out to be.

Aimed at exploring children's perceptions about themselves and
others, the tasks were simple, mirroring the kinds of activities which
the children did daily in the classroom. They included painting portraits
of the self and of others, writing stories collectively about making friends
and drawing pictures of playtime activities. Other tasks – I did not
dignify them in my mind with the term 'research instruments' – were
designed to elicit more detailed information about the patterning of
friendship and were akin to traditional sociometric techniques. Used
by social scientists since at least the 1930s, these research techniques
yield, in pictorial form, the networked pattern of friends within a group.
Children are asked to name the names of those who are their friends
and, through plotting naming patterns in graphic form, a network of
friendships can be literally drawn out.

But there has been a growing realisation that sociometric techniques
do not quite deliver the goods (Hallinan 1981; Mannarino 1980).
Various refinements have been made so that the techniques might yield
more accurate data about social isolates in a group or provide more
reliable indications of day-to-day friendship behaviour (Denscombe
et al. 1986). Elsewhere (James 1993) I offer a fuller critique of their
use in understanding children's friendships. Suffice here to note that
sociometric techniques obscure a number of important issues. Two
are worth reiterating: first, those whom children say are their friends
are not necessarily those with whom they play regularly; second,
those who children do not name may yet be the friends in whom they
most often confide.

What then did I learn from the sociometric data I gathered? The
short answer is very little more than I already knew and a good deal
less than that which the children's teachers had known for some time.
In most instances, although not all, I could have predicted the pattern
of friendships which I laboriously later drew out. However, although
the data on who likes who and who doesn't contained few surprises,
undertaking the exercises was illuminating. It confirmed my understanding of children's social relationships through providing me with
evidence of the processes of friendship making and the meanings
which friendship has for children. Moreover, before my eyes I could
see the tensions which structure childhood as a collective culture

being teased out by the children, both metaphorically and literally, as they went about doing the sociometric tasks they had been set. I was given confirmation of the ways in which power operates between children and the differential wielding of power by girls and by boys through observing with a growing disquiet the struggle for friendship and companionship which often follows the playing out of marginalising strategies amongst a group of children. And, by turn, in the solitude of my study I was led later to reflect on the structuring of that categorical culture of childhood by members of the cultural category of children through, in part, reflecting on my own, as well as their, discomfort. Most recently, taking up some of Mary Douglas's (1982) observations on grid and group, I have come to a more detached reflection upon the process of being friends, by exploring the kinds of hierarchies and boundaries which structure the cultural experiences of children, providing relatively closed or open access to different social environments and facilitating multiple strategies of social action (James and Prout 1994). Thus what had been conceived expediently as a strategy to underscore my status as a researcher not only yielded exceedingly rich insights, but also set in train a series of intellectual challenges in relation to both theory and methodological practices.

My 'fieldwork moment' was, then, no blinding flash of insight nor magical meeting of minds; instead, more prosaically, it was a further shuffling of the pack of cards which produced, finally, a full house. In the following reanalysis of three pertinent fieldwork moments I lay out, therefore, the process of anthropological interpretation from initial instinctive reactions to later, more considered, reflections, to show how the subjective experience of such 'moments' feeds into the later, more academic, interpretation of culture (Geertz 1975). This chapter is not therefore simply a restatement of the potential of inductive theorising; rather, for me, it represents an attempt to unravel the complexity of the process of interpretation, a way of filling in the implicit stages of theorising, characterised by Geertz as the shift from that 'awkward fumbling for the most elementary understanding to a supported claim that one has achieved that and surpassed it' (1975:25).

ACT 1: CONTEXTS AND CONTEXTUALISING

I began tentatively to try out my version of sociometry first amongst a group of six- and seven-year-old children and later with those aged

eight and nine. I was keen to explore what friendship meant to young
children in the light of expressed concerns about their frailty when
compared to those of adults (Bigelow and La Gaipa 1980) and contemporary disquiet about bullying in schools (Tattum and Lane 1989).
Is friendship conceived as an affectionate and enduring relationship
or a more fleeting and utilitarian engagement of one child with
another? Is there just one idea of friendship to which all children aspire
which palls in comparison with other types of social relations, as
developmental psychology would suggest? Or are children, on the
contrary, sophisticated and shrewd operators in their day-to-day
encounters in the playground? Do they engage in a variety of types
of friendships with different children, demanding mutual trust from
some whilst enjoying a more simple companionship with others? Are
boys' friendships different in kind from those of girls or, is it simply
a difference in style which makes their voluble and visible public demonstration of 'being friends' such a stark contrast to the closely intimate
friendships of girls?

That these are not simply empirical matters, amenable to testing veri-
fication, but instead complex issues of definition and interpretation,
arises from the fact that friendship is culturally and contextually variable
(Paine 1974; Allan 1989). Indeed, it was endeavouring to grasp the
shifting and slippery quality of childhood friendships that prompted
my development of the particular research technique which turned
out to be so thought-provoking: the friendship book.

For each of the six- and seven-year-old children I made a simple
booklet in which was inscribed four statements about friendship
feelings and actions which the children had to complete: 'I like to sit
by ... '; My friends are ... '; I like to play with ... '; ' ... is my best
friend'. Aware of the criticisms levelled at standard sociometric tests
which ask children simply to name the names of four or five friends,
using these phrases I aimed to explore a more subtle range of social
contacts. No limit was set on the number of names. The children
completed the first two sentences and then, about six weeks later,
finished the task. Heeding the caution that 'children's choices of their
best friends are at least partially – if not substantially – affected by
momentary factors like beside whom they are standing or sitting when
they complete the test', I sat alongside the children as they carried out
the task (Foot et al. 1980:3). When requested, I gave assistance with
spelling or wrote down the names told to me with the intention of

eliminating any purely practical difficulties associated with the act of writing. I did not wish any child's list of names to be artificially cut short by writer's cramp or fear of failure. Thus, sitting side by side on two small chairs, we would jointly survey the class. And it was from this vantage point that I saw the power which naming names can have.

Some of the six- and seven-year-old children completed the task with relative ease, hurriedly scribbling down names called readily to mind. Six weeks later, these same names reappeared, with one or two exceptions, and later analysis revealed there to be many reciprocal nominations. Bill appeared in John's list along with Sam and Sam cited both John and Bill as his friends. Clearly such children had relationships which fulfilled certain recognisable criteria for 'being friends': their relationship was enduring and mutually rewarding, at least in terms of companionship if not affection. The ease with which these children recalled the names reflected, I felt at the time, the surety of those friendships.

However, other children seemed to find the task more difficult. Perhaps it had no interest for them? Perhaps, for children without close friends, it was a rather delicate subject? Such were my thoughts as some children agreed to participate only if I wrote down the names, whilst others told me, by way of explanation, that these were the 'friends' with whom they had played that morning. Analysing their responses later it was clear that, for these children, friendship was a rather more temporary affair: their second lists contained few repeated names.

A few children were more impassive and detached in their listing. Obediently, at my request, they recorded the names of a number of children in their booklets. Glancing over their shoulders at their lists of nominees I saw – somewhat to my surprise – that many were central players in the classroom networks, often those who demonstrated particularly good social skills – they were social competents, exponents of the art of sociality (James 1993), but not, however, children with whom the passive children ever played. This discrepancy presented me with a first conundrum: did this simply reveal the inadequacy of sociometric techniques or, more worryingly, the naivety of my own particular version; did it reveal these particular children's psychological immaturity in relation to the task set; or, was there something to be gleaned about the act of naming itself? Carol's actions were a case in point. At first she seemed unsure about which names to write in her booklet. Perched on her chair, she surveyed the class and sucked her pencil end. Intermittently, she jotted down a name or two and then looked about again as if to see, from those faces in the room, whom her friends might be. After a while she finished and handed me the booklet remarking, with an air of surprised satisfaction, 'I've got lots of friends.' Seemingly, it was for her a pleasurable experience to see such a list of 'friends' written down.

For me, however, it was disturbing. It made defining and explaining children's friendships a much more difficult task: what did 'friendship' mean if 'friends' could be those with whom one had but a rare and fleeting contact? I felt a flash of irritation. Had setting Carol the task been a waste of time? A second more compassionate thought struck me. I knew that Carol seldom had a companion at playtime and that many of the other girls regarded her, at best, as a hanger-on but, at worst, as a nuisance in their affairs. Becoming increasingly unhappy on Carol's behalf, I was certain that Carol was aware of the patterning of friendships and that she knew it was a network in which she herself played no part. Why then did she list these names?

At that moment in the classroom, I felt exposed in my role as a researcher, a role which usually in my daily life at school remained a hazy aspect of my identity. Through my use of a research tool had I unwittingly drawn Carol's attention to her own social isolation or had I simply found confirmatory evidence of her lack of social skills? Or was it, perhaps, that through asking her to draw up a list of names, Carol momentarily experienced that sense of belonging which she was unable to do on a daily basis? In her imagination had she perhaps traversed that separating boundary and found a place in the hierarchy of prestige from which she was normally excluded? Was that why she seemed so pleased?

The latter interpretation seemed to be the most satisfactory (see James 1993:220). However, time for further reflection allows me to cast a slightly different light on Carol's actions. In an earlier piece of work I have shown how the nicknaming practices of children chart out their social networks, sorting out the style-setters from those who are simply followers and those identified as outsiders (James 1979). In Carol's recording of names, I can now see a similar process of social classification. The names she recorded were the names – not of 'my friends', that is, Carol's friends – but of those children who fall into a more inclusive category of 'friends', that is the category of 'children who have and are friends'. These are the visibly popular children, the children known to be central to the network of social relationships which, in any particular locality, structures the culture of childhood. These are the children skilful in the art of sociality, those who, in effect, are the most effective members of the category 'child' and who maintain a strong position in classroom and/or playground hierarchies.

Now distant from the emotional tenor of the encounter where I was drawn by sympathy to engage and deal directly with Carol's social marginality, it is thus possible to offer three interpretations of her actions, interpretations which I argue are compatible, rather than mutually exclusive. In listing these names, it might be that Carol, who often liked to talk to me given the fact that she had few others who welcomed her chatter, was showing me that she did have some friends after all. It may also be that, as argued above, she was showing me her knowledge of which girls are the kinds of girls who have friends; and, at the same time, she may, as I initially thought, have derived pleasure from this understanding through pretending to herself that she too could be part of that group, at least on paper, if not in practice.

In thus exploring my interpretive procedure as a shift from subjective involvement with a particular act of writing to a more objective perusal of a list of words distant in time and space from their author, I am not, however, arguing for a parallel progression towards truth. Rather, what this sequence demonstrates is the rounding, rather than refining, of anthropological analysis, what Geertz depicts as 'the guessing at meanings, assessing the guesses, and drawing explanatory conclusions from the better guesses' (1975:20).

Watching seven-year-old Tonia fill in her friendship booklet was also instructive. She engaged eagerly, but secretively, in the task. Shielding her writing with her arms, she quickly scribbled down some names, names which she repeated six weeks later. Subsequent cross-referencing revealed that none of the names she had written down were reciprocated. Her nominations stood alone. Was she simply an outsider? Was she, like Carol, a passive watcher, a commentator on, rather than participant in, the social scene? At the time, I thought this to be the explanation. Now, I prefer to argue that what Tonia's list depicts is the fragile and precarious quality of girls' friendships which, on the surface, seem to be so enduring and intimate.

Tonia had been keen to keep her list secret. It was quickly given to me, to be hidden from prying eyes. The list was short, featuring the names of one particular grouping of girls. These girls had a triadic friendship from which, at that time, Tonia was often marginalised. Though included in their games, she was excluded from their public displays of affection. However, the threesome which Tonia identified as her friends – a friendship between Debbie, Pamela and Ellie – was often marred by arguments and disputes. Fieldnotes recall accusations of bossiness between Debbie and Ellie, of Pamela 'telling off' Debbie, of Ellie being spurned by Pamela and my own feelings at being often used by the girls as a repository for tales of their squabbles. As they argued, teased and embraced each other, so the bonds which tied them threatened constantly to dissolve. In writing down their names in her friendship booklet, Tonia was, I thought, staking out her claim to be included should the triad irrevocably fracture (James 1993). Often alone, yet possessed of self-confidence, this list seemed to me to express her desire to belong.

Further reflection inclines me now to a slightly different view. My memory of Tonia is of her aloof and self-contained demeanour, her confidence in talking to me about other children and her silence about her own difficulties at home. This personal style contrasted vividly with the bustling, often spiteful gregariousness of the other girls, those whom she listed as friends. It seems possible, therefore that what Tonia was also offering me was her considered judgement of the power struggles taking place so visibly between Debbie, Pamela and Ellie. Sociometric analysis of these girls' booklets later revealed that whilst each acknowledged the others, they also looked for friendship outside the triad, nominating as friends girls in other close friendship relations. Thus, I would argue that Tonia's shrewd reading of the social context was not simply (or only) a yearning to belong, to be part of a group and to have her social presence acknowledged; it registers her astute awareness of the more general tensions which structure children's culture that create boundaries which include in the moment of exclusion. As I have described elsewhere (James 1993), the tensions between indi- viduality and conformity, between equality and hierarchy create a delicate framework around which children's identities are strung. In her claiming of a future friendship Tonia has made me aware of the precarious nature of those identities. She showed me that they are identities at risk, mere tentative statements of belonging.

ACT 2: TEASING AND TEASING OUT

To see childhood identities as contingent and provisional is perhaps fitting for those who have yet to be deemed social by the adult world. Yet, in the clamour and tumult of the playground, this seems incongruous, so vibrant are the personalities and selves on display. None the less, the tense enactment of friendship relations amongst a group of nine-year-old girls in the summer term of my year at school nudges me further towards this interpretation. At the time, however, it seemed simply a bizarre and emotionally charged event and one which I, with growing anxiety, realised I had unwittingly unleashed.

In a class of nine-year-old children who, in a few weeks time, would be leaving for the more grown-up world of a middle school, I set out to conduct more sociometric tasks. I used a similar strategy, asking the children to name those who they liked to be with and to identify any best friends. However, rather than conducting the task individually in relative privacy, the class completed it as part of a lesson. All was proceeding smoothly when some girls, dominant members of the class, let it be known whom they had chosen for the 'friends' and 'best friend' categories: 'I've put you down', 'Have you put Susie down?', and so on. This sparked off a chain reaction. Those who had been included on another's list hurriedly added that girl's name to their own, whilst those who had been omitted crossly scribbled out the namers' names. New pieces of paper were urgently sought, rubbers demanded and hands carefully concealed lists of names. After a while a more direct interrogation began to take place: 'Did you put me down?', 'Who did you write down?' Answers were knowingly withheld; insinuations made; hints dropped; past indiscretions remembered. The emotional temperature ran high. I watched with an increasing nervousness. Then, one girl hit upon a solution. For the category 'best friend' she created an acronym using the initial letters of all her friends: A,C,L,H,E,N,S. News of this device quickly spread and the threat of giving offence was dissipated: a whole group of girls could, metaphorically, become one girl. A,C,L,H,E,N,S was a best friend to all. Impotent to stop its inexorable course, it was with relief that I joined the girls in their enthusiastic endorsement of the acronym. Distressed and perturbed by the event which I had inadvertently caused, I was temporarily at a loss to explain it.

Although for the girls this incident seemed soon forgotten – A,C,L,H,E,N,S was never mentioned again – for me it was a seminal experience. Tired by the emotional rigours of my fieldwork, by constant attention to words and actions which in ordinary life constitute the muted backdrop for more extraordinary events, I was keenly attuned to the changing tenor of children's relationships. Each verbal slight, teasing comment or sly punch was for me further evidence of the multitude of ways in which social distance and discriminating difference are subtly woven into the fabric of ordinary social life. Bearing the emotional burden of so many different moments of exclusion or teasing, experienced by so many different children during the year, I had become acutely sensitive to the most subtle nuances of speech and behaviour. Thus, although the invention of A,C,L,H,E,N,S was only strategically important for the girls for that one afternoon, for me it was a powerful artifice in seeming to encapsulate the power relations through which children's experiences of childhood are mediated.

In their games (James 1993), girls allow or forbid others to play and some girls, but not all, are recognised authorities on the form and process of playing games. But such girls know how to play through knowing about more subtle power-games, I would suggest, games such as the one that I witnessed that summer's afternoon.

That girls' games are games played by two or three, unlike the team games played by boys, is well documented (Lever 1976) and, just as well known, is that girls' friendships are more excluding than those of boys. Girls often find themselves in close paired relationships, rather than as members of a more inclusive gang or team (Foot et al. 1980). That each mode of being may facilitate the other has not, however, been remarked. What my sociometric task did was to highlight the parallels in behavioural style between these arenas. Game ownership – revealed in phrases such as ' She won't let me play' or 'Can I play?' – mirrors the having and taking away of friends which is so charac-teristic of girls' social relationships and which, on that afternoon, was promising to take place. The threat to exclude from a game parallels the threat to take away friendship. Thus, the pattern of being-in-the-world learnt in one domain becomes a template for action in another.

Recognising this has been pivotal to my understanding of the culture of childhood and informs how I can now write about childhood as a cultural context (Geertz 1975). I have observed elsewhere (James 1993:100) that hierarchical relationships and those stressing equality are, although opposites, clear features of the ways in which children relate one to another, creating a tension which has to be finely balanced by children in their social relations. Reworking this idea more recently, in relation to concepts drawn from cultural theory (Douglas 1982), it is possible to depict these tensions as examples of the different and shifting social environments between which children must move in and through time, environments framed by hierarchies and boundaries (James and Prout 1994).

For example, during the day a child might move between the rigid environment of the classroom and the more consocial world of a youth club, or experience the contrasts between a competitive and hierarchical peer group culture at school and a caring, more equal relationship with friends in the neighbourhood. Successful – that is, socially skilled – children are those who have learnt the strategic flexibility necessary to act across these changing contexts. Those who have not may find themselves ignored, marginalised or more actively excluded. In particular contexts, they may be unable to negotiate the patterned relationships through which identities are articulated.

Returning to that fieldwork moment, and reflecting on it from this new perspective, I can see now what the creation of the mythical A,C,L,H,E,N,S achieved. It was not simply a symbolic gesture of reconciliation. Rather, it was a restoration of the balance of power by a highly skilled social actress. The concept of 'best friend' implies a competitive hierarchy, which potentially excludes, whilst that of 'friends' stresses more equal and inclusive relations. What the sociometric task did was to potentially upset the fine balance of the girls' relations with one another which, possibly with the prospect of soon moving schools, was at that time under a particular strain. My research demand that names be assigned to these different friendship environments threatened to throw the emotional see-saw off balance. The inventor of A,C,L,H,E,N,S, sensing potential disruptive disharmony, found a strategy to avoid a permanent, or damaging, change in the pattern of classroom relations.

ACT 3: REFLEX AND REFLEXIVITY

My last fieldwork moment addresses the question of boys' friendships. It was for me perhaps the most poignant episode in my fieldwork, most of all because I did not realise that it had even taken place until some time later. I am back in the class of six- and seven-year-old children and sitting with Matthew who is writing out his list of names into his friendship booklet. Matthew's list, like Carol's, is long and varied and, again like Carol's, includes the names of popular children, alongside that of his teacher and children from outside the school. I know, however, that Matthew has no firm friends and, moreover, that he is often scapegoated by the other boys. I puzzle over his list as he writes it down. Unlike the other boys who would laugh and tease me, sometimes chatting and rarely confiding, Matthew kept his distance but, occasionally he would engage me in conversation. With this list perhaps, like Carol, he is making a bid to belong or trying to demonstrate his popularity to me.

Some months later finds me transcribing the tapes I have made with the children. Bored by this frustrating task I mechanically record their chatter, with only half an ear, looking forward to the time when I can work more creatively with the transcript. Suddenly, my attention is caught by a lone voice breaking into the conversation I am busily noting down. The talk among a group of boys is of playtime and bursting into the conversational flow is Matthew's voice: 'I like Charles.' Later, Matthew can be heard asking me a question: 'Can you tape, I like Charles?' I reply, 'It's taping everything.' And, later still, Matthew's voice chimes in for a third time: 'I like Charles.'

What struck me was that the other boys simply did not hear him. He might as well have not been there. That I had heard his plea just once, and then answered him somewhat testily, preferring to listen to the chatter of the other boys, chilled me. Had I too effectively obliterated him from the social map? Had I colluded with the other boys in excluding Matthew? I could not remember him attempting to join in the conversation. Only the tape stood testimony to his efforts and to his failure.

At the time of the recording Matthew had been anxious to find a friend in Charles but to no avail. Charles actively rebuffed him and Matthew himself knew that he had no friends. Later he was to draw me a desolate picture of playtime which imaged two tiny stick figures crouching at the foot of a black and lowering building with a dark sun seemingly compounding their predicament. That Matthew knew he was an outsider but did not know how to change the situation was abundantly clear.

Reflecting on this after I had listened to the tape, I interpreted these interruptions, initially, as Matthew's attempt to register a bid for friendship, as demonstrating his desire to belong to the central grouping of boys (James 1993:220–1). But I think more can be said. That the other boys appeared simply not to hear Matthew, rather than that they chose to ignore or snub him is the most telling aspect of this event. What it underlines is Matthew's lack of social skills. It shows clearly that Matthew did not know how to belong, that Matthew had no style (cf. James 1986).

This concept of style, which previously I have used to describe the ways in which older boys and girls socialise with one another, can be situated more theoretically in terms of the second set of tensions which structure the culture of childhood. These I have described as the tensions between conformity and individuality (James 1993:100). A skilful child manages to be one of the crowd whilst avoiding being submerged within it; he or she is an individualist who is neither an eccentric nor a passive follower of fashion. It was this balance that Matthew was unable to achieve which meant that, quite literally, the other boys did not recognise his bids for friendship. He was a non-person. Indeed, he only impacted upon their lives when, by misbehaving, he presented them with a model of what a boy should not be. Matthew, therefore, did not feature in the social landscape except as a stereotypical other, against which other boys measured their own conformity or as a useful scapegoat for individual miscreant acts. Moreover Matthew's social lack was the other children's social gain: his difference sharpened up the boundaries of belonging.

That boys' games are commonly team games (Lever 1976) mirrors the emphasis which boys place on belonging to a group with respect to friendship and social relations. Within any huddle of boys, bragging and boasting in the playground, will be recognised leaders, known masters of style whom the others admire and may try to mimic. Ironically, however, it is through a strong conformity that these individualists gain their admirable reputations. Such boys, for example, although championing the masculine qualities of toughness and physical strength in their verbal accounts of deeds of derring-do, of battles won and lost or yet to be fought, are not the boys who get caught in playground brawls or who gain reputations as bullies. They are instead strategic players, who find a flexible response to the subtle shifts in the social and emotional contexts in which they find themselves. They may dally with the bounds of good behaviour but do not overstep the boundaries set for belonging.

CONCLUSION

I have talked of 'fieldwork moments' quite deliberately. The 'moment' is an apt metaphor for the processes I have been depicting. In the physical world, the leverages and tensions involved in achieving balance are measured in moments and, for me, the moments I have been describing starkly revealed the fulcrums on which children's power relationships are poised. Being able to match the demand for conformity with a display of individuality, to participate in competitive and hierarchical relations while professing equality with one's friends is a difficult balance for children to achieve, and yet it is that towards which they must strive if they wish to belong. But these moments also underline the importance of ethnography as a method for researching children and childhood. As noted elsewhere (James and Prout 1990), it is only relatively recently that an ethnographic approach – and participant observation in particular – has come to be regarded as a key methodological tool to access children's perspectives on the social world and, in this final section, I reflect further on its value.

First, from the examples discussed above, it is clear that what participant observation makes possible is the exploitation and later explication of the serendipitous, of the encounter that could not have been imagined, of the questions that could not have been voiced. Commonly acknowledged now as a strength, rather than a weakness – the encounter described by Geertz (1975) in his essay on the Balinese cockfight being one of the most instructive examples – in work with children it is, I would argue, essential. The positioning of children as vulnerable beings, not yet fully formed nor socialised in the ways of the world (Jenks 1996), has meant that adult interpretations of what children say and do have been subject to a framing of incompetence, a judgement sustained by the power and prevalence of the notion of the developing child (James et al. 1998). It is a framing which has hitherto thus predefined many of the key issues in childhood research, primarily in terms of the problems which children present to the adult world (for example, in terms of their socialisation or failure to thrive) or the problems which adults judge children to experience (for example, educational failure or lack of friends). Those issues which children themselves find important or particular to their social lives have, in this way, often been marginalised. What participant observation permits the ethnographer to do, however, is to question such an adultist bias through seeing the world more through children's eyes and experiencing the everydayness and detail of their social lives. And, as in the adult world, as Geertz observed, it is often the case that 'small facts speak to large issues' (1975:23).

In learning to understand children's friendships, for example, it was observing Carol's contemplative sucking of the end of her pencil and Tonia's secret shielding of a list of names which alerted me to the power which the naming of names has for children. Sociometric methods, with emphasis placed on formal questions and the ranking of lists of names, would not permit the observation of such actions which, for the ethnographer, enable the patterning of children's social relations to be revealed as they slowly unfold through a series of seemingly inconsequential and commonplace actions and events over a period of time.

Central to this, as Hastrup notes, is the anthropological imagination which 'makes room for novel connections that come out of experience' (1995:63). As Geertz observes (1975), anthropological understanding is always of a third order. Events which occur during fieldwork – often literally in a moment – become transformed. Through the process of recording – the note, the diary, the tape transcript – a moment is taken out of time and space. It becomes frozen, extracted as an object to be regarded, observed and analysed at leisure in another time, in another mood (Rapport 1991). So it is, through this freeze-framing, that the subtle nuances of social life come to be displayed, replayed and relayed as anthropological insights or explanations.

A second and related point concerns the experiential aspects of participant observation. The moment of action is subject to more than the ethnographers's gaze. As Stoller (1989) has observed, all the senses become engaged and it is precisely the third-order nature of anthropological interpretations which enables us later to make sense of those sensory experiences. In the events described above, the complexity of emotional engagement with the 'field' is detailed in the moment of its occurrence during fieldwork with the children and during the later phase of analysis. In both instances a reinterpretation is offered, a new understanding gained through the passage of time, away from the heat of the moment. What this points to then is that, as Geertz wryly observed, cultural analysis is not about the 'conceptual manipulation of discovered facts', a systematic and objective collation of data. It is a more messy yet more insightful process which involves 'guessing at meanings, assessing the guesses, and drawing explanatory conclusions from the better guesses' (1975:20). Narrated above is an account of this guess-work.

What makes me feel in awe of children, therefore, is their capacity to gain the insight that they do into the child's world which they inhabit. For them there is not the luxury of a period of quiet and lengthy contemplation. Instead, such moments as I have described in detail here, and thought about for considerably longer, take place in a flash, constituting but fleeting lessons in the operation of power. They offer mere glimpses of what friendship might entail, tiny gestures of disapproval or acclaim. It is this struggle for effective participation which makes the term 'socialisation' seem such a woefully inadequate description. It is this which also suggests that anthropologists should not study childhood as the site for the socialisation of children, but as that moment in the life course when both agents and styles of agency are being formed.[2]

NOTES

1. An earlier version of this chapter appears in *Childhood* 3(3): 313–30. It appears here with permission.
2. D. H. M. Brooks was my supervisor at Durham University. This chapter celebrates his inspiration and dedication as a teacher. *In memoriam* 29 April 1994.

REFERENCES

Allan, G. (1989) *Friendship: Developing a Sociological Perspective*, London: Harvester Wheatsheaf.

Bigelow, B. J. and La Gaipa, J. J. (1980) 'The development of friendship value and choice', in Foot et al. (1980).

Bourdieu, P. (1977) *Outline of a Theory of Practice*, Cambridge: Cambridge University Press.

Denscombe, M. et al. (1986) 'Ethnicity and friendship: the contrast between sociometric research and fieldwork observation in primary school classrooms', *British Educational Research Journal*, 12 (3): 221–35.

Douglas, M. (ed.) (1982) *Essays in the Sociology of Perception*, London: Routledge and Kegan Paul.

Ennew, J. (1986) *The Sexual Exploitation of Children*, Cambridge: Polity Press.

Foot, H. C., Chapman, A. D. and Smith, J. R. (eds) (1980) *Friendship and Social Relations in Children*, London: John Wiley.

Geertz, C. (1975) *The Interpretation of Cultures*, London: Hutchinson and Co. Ltd.

Hardman, C. (1973) 'Can there be an anthropology of children?', *Journal of the Anthropology Society of Oxford*, 4 (1): 85–99.

Hallinan, M. (1981) 'Recent advances in sociometry', in Asher, S. R. and Gottman, J. M. (eds) *The Development of Children's Friendships*, Cambridge: Cambridge University Press.

Hastrup, K. (1995) *A Passage to Anthropology*, London: Routledge.

Holt, J. (1975) *Escape From Childhood*, Harmondsworth: Penguin.

James, A. (1979) 'The game of the name: nicknames in the child's world', *New Society*, 14 June.

James, A. (1986) 'Learning to Belong: the Boundaries of Adolescence', in Cohen, A. P. (ed.) *Symbolising Boundaries*, Manchester: Manchester University Press.

James, A. (1993) *Childhood Identities*, Edinburgh: Edinburgh University Press.

James, A. and Jenks, C. (1996) 'Public Perceptions of Childhood Criminality', *British Journal of Sociology*, 47 (2): 315–31.

James, A., Jenks, C. and Prout, A. (1998) *Theorizing Childhood*, Cambridge: Polity Press.

James, A. and Prout, A. (eds) (1990) *Constructing and Reconstructing Childhood*, Basingstoke: Falmer Press.

James, A. and Prout, A. (1994) 'Hierarchy, Boundary and Agency: towards a theoretical perspective on childhood', in Ambert, A. (ed.) *Sociological Studies of Childhood*, 7: 77–101.

Jenks, C. (1996) *Childhood*, London: Routledge.

Lever, J. (1976) 'Sex differences in the games children play', *Social Problems*, 23: 478–87.

Mannarino, A. P. (1980) 'The development of children's friendships', in Foot et al. (1980).

Okely, J. (1992) 'Anthropology and autobiography: participatory experience and embodied knowledge', in Callaway, H. and Okely, J. (eds) *Anthropology and Autobiography*, London: Routledge.

Paine, R. (1974) 'Anthropological approaches to friendship', in Leyton, E. (ed.) *The Compact: selected dimensions of friendship*, Newfoundland: University of Toronto Press.

Prout, A. (1989) 'Sickness as a dominant symbol in life course transitions: an illustrated theoretical framework', *Sociology of Health and Illness*, 4 (11): 336–59.

Rapport, N. (1991) 'Writing fieldnotes: the conventionalities of note-taking in taking note in the field', *Anthropology Today*, 7 (1): 10–13.

Rapport, N. (1993) personal communication.

Stoller, P. (1989) *The Taste of Ethnographic Things. The Senses in Anthropology*, Philadelphia, PA: University of Pennsylvania Press.

Tattum, D. P. and Lane, D. A. (eds) (1989) *Bullying in Schools*, Stoke on Trent: Trentham Books.

Woods, P. (1987) 'Becoming a junior: pupil development following transfer from infants', in Pollard, A. (ed.) *Children and Their Primary Schools*, Lewes: Falmer.

5 THE END IN THE BEGINNING: NEW YEAR AT RIZONG (THE HIMALAYAS)

Anna Grimshaw

PROLOGUE

Henri Cartier-Bresson, the great French photographer, is approaching his ninetieth birthday. The exhibitions which celebrate his life's work give striking expression to a distinctive twentieth-century vision. Cartier-Bresson's photography is distinguished by its structural elegance and profound humanism. His technique is simple. It involves no elaborate equipment or intricate planning. Carrying only a small camera, Cartier-Bresson works rather like an ethnographer, exploring, searching, questioning, guided by his intuitive sense, alert to chance and to the spontaneous. The basis of his photographic approach is known as 'the decisive moment', his remarkable ability to seize a moment from all other possible moments to reveal something new about the world in which we live:

... photography is the simultaneous recognition, in a fraction of a second, of the significance of an event as well as of a precise organisation of forms which give that event its proper expression ... I believe that, through the act of living, the discovery of oneself is made concurrently with the discovery of the world around us, which can mould us. A balance must be established between these two worlds, the one inside us and the one outside us. As the result of a constant reciprocal process, both these worlds come to form a single one. And it is this world that we must communicate. (Cartier-Bresson 1952)

Henri Cartier-Bresson developed as a photographer in the context of the 1930s Parisian avant-garde. He was part of an eclectic collection of poets, artists, musicians, writers and intellectuals. Prominent, too, were a number of ethnographers including Marcel Griaule and Jean

Rouch. 'What you have to understand at this time,' Rouch once explained

> ... in the thirties, 'anthropology' per se did not exist. All the people who were in some way 'artists' or 'anthropologists', well they were philosophers; they were thinkers, they were writers, they were poets, they were architects, they were filmmakers, they were members of only one very wide group. It was, in fact, l'avant-garde. They were exchanging their experiments, and Paris was a kind of strange workshop where there was a sharing of all these experiments. (quoted in DeBouzek 1989:302)

I am not sure that Bill Watson had Cartier-Bresson and the interwar Parisian avant-garde in mind when he proposed the idea of a seminar series based on the notion of an 'important moment' in an ethnographer's experience of fieldwork. For me, however, both are evoked by this notion. Watson uses the example of Clifford Geertz's dramatic entry into Balinese society to draw attention to what I call the 'visionary' impulse of anthropological work.[1] By this I mean that he refers to the transformative experience of fieldwork – that split-second when something unexpected and remarkable happens which changes not only one's perception of the world but one's own perception of self. Specifically, the notion of the 'important' or 'decisive' moment challenges the assumption that anthropological understanding involves the progressive accumulation of knowledge; rather, as Watson implies, a dramatic shift in levels of engagement can occur, transforming the spectator into a 'seer' (Stoller 1989:40). Such a conception of ethnographic enquiry immediately prompts questions concerning the nature and status of certain fieldwork experiences. What are these important or decisive moments? Can they be described? Do all ethnographers experience them? Can they be accommodated within the conventional narratives of ethnographic writing? What do they reveal about anthropology's claims to be an intellectual discipline concerned with description and analysis? Are transformative moments admissible within contemporary anthropological discourse or are they indeed evidence of some kind of ethnographic magic? (Stocking 1992).

All of these questions are, I believe, acutely posed in the case of ethnographers trained by Malinowski and Radcliffe-Brown in the classic tradition of British structural-functionalism. For the emergence of this distinctive school was not marked by the same innovative spirit which is so tangible in Rouch's characterisation of Parisian intellectual life during the 1930s. There is little evidence of a creative workshop in which ideas and practices were exchanged or anthropology's visionary, transformative moments were celebrated. It might be argued that the opposite tendencies were at work in interwar British anthropology. Historians of the emergent British school (for example, Kuklick 1991) draw attention to a process of consolidation not of experimentation. The desire of its leading figures was not to shock but rather to win acceptance from established intellectuals for their new kind of scientific enquiry. There was a concern to separate and clarify key concepts and categories in the drive to establish scientific ethnography as the paradigm upon which claims to professional recognition could be built. Most notable among these were the distinctions between truth and fiction, science and the subjective. Such dualities became enshrined in the theory and practice of postwar academic anthropology. They remained largely unexplored and unquestioned until the mid–1980s.

Given this particular history of anthropology as an academic discipline in Britain, it is perhaps not surprising that the successors to Malinowski and Radcliffe-Brown have manifested considerable scepticism about recent reflexive work.[2] As Watson notes in his Introduction, expressions of interest in autobiography, the subjective, speculative or experimental writing and so on are often greeted with impatience by those eager to press on with the collection and analysis of ethnographic data. Although at some level there is now an acknowledgement of the legitimacy of critical self-reflection, I, like many others, remain frustrated by the continuing resistance to any serious engagement with issues raised by the discipline's moment of self-consciousness. (Notable exceptions to the predominant trend include, of course, Okely and Callaway (1992) and Rapport (1993, 1994).)

I myself was trained in the tradition of structural-functionalism. From my own experiences as an ethnographer, I know only too well the contradictions at the heart of the enterprise. Initially, as this chapter will reveal, I sought a certain resolution to questions concerning what Okely calls 'the self and scientism' (1975). Now I am more interested in drawing creative energy from the contradictions themselves. I flourish in anthropology's liminal spaces. By exploring those marginal areas where conventional categories are confused and difficult to sustain, I frequently encounter exasperation from colleagues anxious to draw lines between what is fiction and what is ethnography, what is imagined and what is real, what is remembered and what is witnessed, what is subjectively felt and what is objectively described.

Any investigation of these areas, however, raises the additional problem of form. For what may be discovered here is by its very nature resistant to incorporation within the discourse of academic anthropology. One is forced to experiment or improvise with form in the attempt to express new kinds of ethnographic understanding. But this, too, is fraught with difficulties. Such attempts are often dismissed as not 'real' anthropology, or they provoke a critical response which calls for the reintegration of the experimental approach within the familiar conventions of intellectual exchange. Hence one is called upon to explain, to argue, to contextualise one's work in relation to a professionally legitimated field. It is deeply frustrating because, of course, the innovation is driven by a desire to subvert those very constraints and to challenge the conventional expectations of an academic audience.

This chapter is an attempt to find a creative synthesis of form and content. Its construction, the different parts and narrative movements, has been shaped with my substantive concerns in mind. I begin with an epigraph from T.S. Eliot's *Four Quartets*. My purpose in opening with this poetic image is to orient the reader in a particular way towards what follows. For, although my chapter appears to develop through linear time, containing as it does a beginning and an end, it also contains another movement, one which is circular and folds back upon itself in a recursive rhythm. It evokes, I believe, the process by which we attempt to find meaning in fieldwork experience.

In his poem 'Little Gidding' Eliot describes how, at the end of what he calls exploration, we do not stop searching and that paradoxically where we end is where we began, only with the difference that this time we perceive that original starting point in a new light. What Eliot is arguing here, it seems to me, is that the active creative life is forever engaging in a cycle of experiential quests which lead back to a reconsideration and reevaluation of earlier judgements, earlier ways of seeing things. Returning to one's memories of fieldwork is like that.

TEXT AND CONTEXT

I will present two extracts from my book, *Servants of the Buddha*, as the starting point for an exploration of what might constitute 'decisive' moments in ethnographic fieldwork. The first passage is set in the temple of Rizong monastery; the second describes an event which took place on the mountainside behind Julichang, the nunnery attached to Rizong monastery and the place where I lived. The two passages, focusing upon the arrival of New Year in Ladakh, are arranged in reverse order to their appearance in the book.

I tasted the sweet fragrance of the incense as I joined the nuns inside the temple. I looked around nervously, awed by the splendour of the occasion and the ostentatious display of Rizong's wealth and power. The doors of the temple were closed behind us. The night dissolved into memory and we entered another world.

The monks and nuns prostrated themselves before the great bronze figure, and began to repeat their adherence to the Buddha, the doctrine and the monastic community. Their voices hardly rose above a whisper, but the chanting had a momentum of its own, bridging the disjunction of time.

I lost my sense of duration in the temple. I was adrift, only conscious of the uncertainty of my perceptions. The light played tricks. Its penetration was refracted and scattered through the thickening haze of incense; its brilliance transformed into a deep hue as it reflected the rich golden silk worn by the monks. The soft murmur of faith enveloped me. It was dreamlike, elusive and strange.

The stinging cold air on my cheeks eventually roused me. It was filtering into the temple as the young boys drew back the heavy doors. The pale morning sky cast shadows through the interior. Dawn had broken and with it we had passed from the old year to the new. (Grimshaw 1992:99–100)

Suddenly, the women shattered the stillness of the winter day. They began to call out the names of former inhabitants of Julichang, nuns now deceased. Their voices were shrill and brittle, their chanting broken by wails and sobs. It was a violent emotion which echoed eerily down the long, dark valley.

I stood awkwardly, an onlooker at a scene which was frozen in the chill of the Himalayan air. It was one I could hardly grasp. I was familiar with the physical strain of the women's lives, their resilience and stoicism; but I had not anticipated this – an uninhibited expression of pain and grief.

... I perceived a frightening void, a chasm of bleakness and desolation. It was as if it had opened up before them and they hovered at its brink. Slowly their sobs subsided and the women regained their composure. They quickly covered the cracks which only a moment ago had laid bare their vulnerability. The outburst had left them defensive and self-conscious.

We sat on the cold, rocky ground and each of us pulled out a bowl from the front pouch of our woollen robes. Sonam gave us some of the meat and onions, followed by a small quantity of beer. No one spoke; but we cast nervous glances at one another as we consumed them. Tsultim divided the rest of the offerings. These, too, we ate; but a small amount of each was left on the three stones. The pots containing the remainder of the meat and beer stood nearby.

Chilled and pale, the nuns struggled to their feet. We were unsteady and our frozen limbs moved clumsily. We nearly laughed aloud but checked ourselves, fearful of upsetting the uneasy calm which had returned. Uttering a chorus of shouts and cries, the nuns implored the sleek black crows, hovering a distance away, to swoop down and devour everything we had left on the mountainside. We did not look back as we made our way in silence down the slope to Julichang.

The nuns appeared relieved. It was behind them for another year. They warmed themselves by the fire and prepared to meet the New Year. Their animation was restored and they were again the women whose life and work I shared. But I saw clearly on their blackened faces the traces left by their tears. (Grimshaw 1992:94–95)

The movement from the old year to the new, as with all phases of social transition, is surrounded by a complex series of ritual precepts and practices. In the context of Ladakh, the transition is conceptualised as a fierce struggle between the Buddhist faith and chaotic local forces. For the Tibetan version of Mahayana Buddhism was introduced into this area from the top, percolating downwards from the monastic foundations through different layers of the Ladakhi population until it was eventually incorporated into the lives of the ordinary villagers. The process by which it took root involved an accommodation of many unorthodox lay beliefs and practices by the representatives of monastic Buddhism.[3] The contradictions buried within this history are laid bare at New Year. They are symbolised in the two rituals which I describe in my book – the ceremony performed in the temple at Rizong, symbolising the annual triumph of Buddhist orthodoxy, is inverted at Julichang by the unexpected irruption of a strange unknown power. If my experience of the first ritual occasion was one of completeness, it was the scene on the mountainside above the nunnery which lodged most forcefully in my memory.

The nuns, with whom I lived and worked, played an important role as mediators between monastic and lay life.[4] They straddled an uneasy divide between spiritual learning and economic drudgery, aspiring through their work to achieve rebirth as monks while being trapped, along with the villagers, in an unremitting routine of productive activity which supported monastic seclusion. During New Year, the pressures of their uneasy position reached a new intensity, shattering the fragile security of our shared world. At this time I, too, was forced to recognise the contradictions embedded in the life I had constructed there. For in this same liminal moment which laid the society bare, revealing its hidden dimensions and fundamental social relationships, I experienced a profound dislocation of my own. And in the moment I perceived the women anew, I recognised that my relationship to them had been transformed.

My choice of a rite of passage, the Ladakhi New Year, is perhaps a conventional place to begin any investigation of what might constitute a decisive moment in fieldwork. Anthropologists have long recognised that the ritual processes of a society can be especially revealing, as normal categories become fluid, boundaries dissolve and the elements of new understanding begin to emerge. But what I want to emphasise here is that my interpretation of such a moment in a Ladakhi Buddhist society is inseparable from my own changing subjective participation in it.

The first extract I presented from *Servants of the Buddha* described my experience of transition as I passed from the old year to the new. This moment, located in Rizong's temple, moved me deeply. The ceremony was a celebration of the spiritual authority of monastic Buddhism, a power enhanced through its concentration within an area of mystical space and one which bound us tightly together as a single community during the passage from one time to another. In slipping from a state of heightened sensory perception into a strange sort of limbo where conventional boundaries of the self dissolved, I had, of course, surrendered myself to the liminality of a classic rite of passage. It was out of time and out of consciousness. In moving through time, I had the sensation of timelessness; through spatial separation, I experienced integration.

The second passage quoted from the book concerned the strange ritual I witnessed being performed by the nuns at Julichang just before New Year. Unlike the enclosed setting of the monastery temple, we were on this occasion outside, exposed on the harsh, rocky mountain slope which rose steeply above the nunnery. The climax was a sudden outburst of uncontrolled, almost wild, power. The past, buried deep in the hidden recesses of the unconscious, intruded violently into the present, and I stood watching, a distanced onlooker to a scene which could not incorporate me. I experienced the moment as one of profound separation from the women whose lives I had begun to believe I shared. Being trapped within real time intensified the sensation. I was conscious, too, of a current pulling me away from that life, one which returned me to an old and familiar terrain – my own past.

The two ritual events associated with the Ladakhi New Year are integral to the book's unfolding narrative. Indeed they constitute its climax. For *Servants of the Buddha* begins with my arrest and expulsion from Ladakh and ends with the same moment; and yet the telling of this episode is subtly altered in the two versions. For example, in the Prologue to the book, I describe how I was originally mistaken for a local woman by the policemen sent to apprehend me. Addressing me in Ladakhi, they asked if I had seen an Englishwoman on the mountain path. Only when they had drawn closer did they realise that I was the person they sought. The description of my arrest which comes at the end of the book, however, makes no mention of this mistaken identity.

I had been living at Julichang without a permit from the local authorities. There was considerable ambiguity about whether the nunnery was situated beyond the security line which demarcated an area prohibited to foreigners. It had been easy enough to travel there. I had simply caught a bus from Ladakh's capital, Leh; from the road I had walked for a couple of hours into the mountains, following the narrow track which ran through the river valley until I reached Julichang. I had decided not to apply for permission to stay in this area for fear of refusal, preferring instead to leave the question open. Repeatedly I sought to push it to the back of my mind; but I continued to remain half-conscious of the issue lurking there, unresolved. The question was eventually forced by Rizong's head monk. He insisted that I notify the police of my presence. Immediately I confronted powerful but conflicting emotions aroused by my fieldwork experience. The relationships and the kind of life I had built for myself at Julichang enabled me from time to time to imagine cutting free from my own past; and yet I also knew that there was an inevitability to my departure. Indeed I suspect that it was the knowledge of departure which made possible my contemplation of its opposite. The arrest when it finally came, however, abruptly resolved the question which I could not resolve in my own mind – whether to leave or stay. The book, too, leaves open this question. But the forced nature of my departure denied an outlet to the confused and volatile emotions surrounding my fieldwork; and for many years I remained deeply troubled by them.

The description of my arrest in the opening pages of *Servants of the Buddha* establishes the event as a violent interruption of my fieldwork, shattering a moment of apparent integration. In the closing pages, however, my arrest appears to offer an escape, a release, from the hidden currents which threaten to overwhelm the ordered life that I had constructed with the women of Julichang. What comes at the beginning of the book then already anticipates its end. 'My life with the nuns of Julichang came to an abrupt end,' I wrote in concluding my Prologue, 'and I was plunged into the bewildering world of Indian politics.' But in the end also lies a new beginning, for, as I wrote in the last lines of the book, 'I began the long walk down the dusty track, through the dark valley and into the wide plains beyond.' Thus, although the narrative begins and ends with the same moment of separation, the meaning attached to it changes in the course of the story. New Year is the crux. The two faces of my arrest, contained in the narrative's beginning and end, are dramatically juxtaposed at its centre in the shape of the two episodes with which I opened this chapter.

My discussion of key moments has not focused upon an investigation of my fieldwork as such; rather I have been concerned with its textual transcription as *Servants of the Buddha*. This seems to me an important question – what is the process by which we attach meaning to experience? For while the experiences themselves are always contemporary, their significance only crystallises later as a consequence of reflection. Thus, although the ritual moments at the monastery and nunnery during New Year are turning points in a narrative of integration/separation, their meaning only emerged in the course of the writing itself. Moreover, the meaning of these moments continues to be modified with each subsequent reading of the book.

Meaning, I suggest, does not inhere in the specific moments of fieldwork themselves, rather it emerges from the patterns of associations in which they become embedded. We are endlessly engaged in the construction and reconstruction of key moments from a place outside those moments. We return to the past of our fieldwork, carrying with us the preoccupations of our present. Habitually, we evoke it as whole, bounded and timeless; but what we hope to find there changes as we move through time. It is this movement, both linear and circular, which I now wish to explore.

I began writing *Servants of the Buddha* about seven years after my fieldwork in Ladakh. Like many anthropologists, I returned from Julichang to an academic setting with an acute sense of dislocation. My unease intensified during the period in which I wrote my doctoral thesis, since the formal demands of that exercise seemed to be seriously at odds with the fieldwork experience itself. I was completing my research work in the period immediately preceding the publication of *Writing Culture* (Clifford and Marcus 1986), the book which perhaps more than any other marked a new phase of self-consciousness in the discipline. Anthropology as I encountered it then was still bound by the set of professional practices which had developed during the classical phase of the British school half a century earlier. As a research student, I was bound by the norms of scientific ethnography. The messy details of fieldwork and the complex, deeply felt relationships with people were to be translated into neat analytical categories and organised into an academic argument. For me, the writing-up process involved a brutal, painful repression of the ethnographic experiences which constituted my fieldwork. The experiences, however, did not go away. Rather I was always conscious of them, hovering like a dark shadow at my shoulder. They were powerful, intense, unexamined, and profoundly threatening to any sense of personal order.

Several years passed. I became involved in other work quite unconnected with academic anthropology. But then a curious and unexpected thing happened. I found myself wanting to write about my life with the women of Julichang. I discussed the matter with Edmund Leach, my former doctoral supervisor, because, although I was ready to write, I did not know how to set about writing an ethnography that would be different from the disembodied, objectifying text of the professional discipline. There seemed to be few alternative models to the conventional monograph. Leach strongly urged me to think of attempting a personal account. For he himself, at the end of his life, had come to embrace an autobiographical approach to anthropological writing (Leach 1989), persuading me that all else was 'fiction'. I decided to try and retrieve letters I had sent from Ladakh to friends and relatives. These documents, along with my diary, I believed would form the basis for the book.

I started to write. Almost immediately I abandoned all the materials I had collected. I became fascinated by what I could see in my mind's eye. When I began to draft the manuscript I had no idea where an exploration of my memories might lead; but I was struck by the vividness of the fieldwork images I was able to summon into my consciousness and by the affective qualities which were attached to them. As I dug deeper, my confidence as a writer grew. There was a creative bedrock of experience to which I had, at last, gained access. Moreover, I sensed that during the years of suppression and anxiety about what lay buried there, a process of sifting and sorting through the chaos of my fieldwork had taken place. This process was unknown to me until I sat down to write.

The first draft of *Servants of the Buddha* was written in the course of a year. The unfolding of the narrative possessed a momentum of its own such that I was not aware of consciously fashioning a coherent story; rather, it seemed to develop, spontaneously, at the time of writing itself. My original points of departure were the intense vivid moments of my fieldwork, scenes such as the two episodes at New Year and my arrest. But to evoke them in my mind's eye was always to trigger other images, creating shifting patterns of association whose meaning then crystallised by being given expression in language. Writing was a process of making sense, of finding meaning in the relationships of elements within a pattern. Fundamentally it was about connecting the present with the past. For I could never recapture those past experiences of fieldwork, I could only recreate them from my position within the present.

THE REFLECTIONS OF A HYSTERIC

'An old photograph in a cheap frame hangs on the wall of the room where I work.' So begins Salman Rushdie's essay, 'Imaginary Homelands':

> It's a picture dating from 1946 of a house into which, at the time of its taking, I had not yet been born. The house is rather peculiar – a three storey gabled affair with tiled roofs and round towers in two corners, each wearing a pointy tiled hat. 'The past is a foreign country', goes the opening sentence of L.P. Hartley's novel *The Go-Between*, 'they do things differently there.' But the photograph tells me to invert this idea; it reminds me that it's my present that is foreign, and that the past is home, albeit a lost home in a lost city in the mists of lost time. (Rushdie 1991:9)

Rushdie's concern here is to explore certain aspects of the migrant experience which lie at the centre of his creative writing. But the questions he raises are, I believe, important to anthropologists seeking to come to terms with fieldwork experiences. For ethnographers are migrants too.

Rushdie is fascinated by dislocations of time and space, and by his own attempts to bridge them, to recover that which has been been lost through the construction of what he calls 'imaginary homelands' (1991:10). While working on his most celebrated novel, *Midnight's Children* (1981), Rushdie was forced to recognise that he could not recover the past of India's independence; rather he was creating it, using his memory to produce a subjective, partial version of history and one he claims to be 'imaginatively true' (1991:10). How the memory works, its tricks, its distortions, its fundamental unreliability become, through his narrator Saleem, central to his exploration of India's modern birth. Thus *Midnight's Children* is not just 'a novel *of* memory', it is also a novel '*about* memory' (1991:10, emphases added). But as Rushdie notes in his essay, a writer who is displaced geographically, and often linguistically, is only experiencing in a more intense way the sense of loss that haunts each of us in our own lives. We are all migrants, migrants from our past.

Rushdie's point of reference, 'the old photograph in a cheap frame', is an important symbol, too, for Terence Davies who seeks through filmic evocation to find a way back into his postwar Liverpool childhood. On the wall of the terraced house hang faded family photographs. Davies places them at the centre of his autobiographical film, *Distant Voices, Still Lives*, the place to which the camera's eye returns, anchoring the self in a past itself recreated as a series of still photographs. And again like Rushdie, Davies is fascinated by how the memory works. His film unfolds as a mosaic of fragments, linked poetically into a whole, and yet each part startling in its intensity, in its luminosity: 'Memory does not move in a linear or chronological way – its pattern is of a circular nature, placing events (not in their "natural" or "real" order) but recalled for their emotional importance' (Davies 1992:74). The past and present, as Davies acknowledges here, can never be linked in a straightforward way, since attempts to reconstruct experience as a linear process in time are always thwarted by the circular movement of the memory. The memory itself is moved by the affective qualities which become attached to certain images and lodge subsconsciously in the mind.

The 'imaginary homelands', which Rushdie recognises the migrant is impelled to create, exist outside of real time and space. They promise wholeness, integration, meaning, to a self stranded in the midst of a fragmented present. But as the work of both Rushdie and Davies reveals, these homelands are flawed, for they are fashioned from scraps, fragments, shards of memory. But it is the very imperfection of the pieces which make them so evocative, so compelling.

At the beginning of his essay, Rushdie remarks that this conception of a homeland is, in fact, an inversion of the conventional model – the past as a foreign country, the past as an unknown but chaotic world which threatens to overwhelm the Freudian hysteric. 'Hysterics suffer mainly from reminiscences', Freud stated with Breuer in their famous 1893 paper on hysteria. These reminiscences, repressed or displaced, are always pricking at the conscious, threatening to burst through and to shatter the fragile order of the self. Freud's great innovation was, of course, psychoanalysis – the development of a language through which these reminiscences could be articulated, ordered, integrated and given narrative form. In this way, the self acquires a history.

Over the last decade or so questions of the self, consciousness, subjectivity and autobiography have come to the fore of anthropological debate. Moreover, writers from outside the discipline are increasingly acknowledged for the insight they offer into processes of ethnographic understanding. Hence Rushdie, Davies and, of course, Freud may now find a place in any investigation of fieldwork experiences. In my own case, the first draft of *Servants of the Buddha* was complete when I came across both Rushdie's essay and Davies's film, *Distant Voices, Still Lives*. Immediately I recognised that this work offered me a context in which I could anchor my own writing. For *Servants of the Buddha* had been written without reference to academic anthropology. I thought that my interest in the 'imaginatively true', in the images in my mind's eye, in the distinctive qualities and movement of memory could only be legitimated by reference to materials outside the domain of anthropology. It was only some years later, after my return to the academic discipline, that I discovered I could in fact find a validation of my approach to fieldwork experiences within anthropology itself. The process by which I made this discovery, however, was unusual. Again it required my engagement with unconventional sources. This time it involved the work of a contemporary novelist, Pat Barker. Just as Rushdie and Davies had enabled me to develop a reflexive consciousness about what I had been doing intuitively, that is, exploring fieldwork experiences through the cracked and distorted lens of memory, so too Barker's trilogy of novels stimulated new questions about the processes of ethnographic understanding.

Anthropology's moment of self-consciousness has had at its core a re-examination of the significance of Malinowski, long recognised to be the key figure in establishing the model of modern fieldwork-based ethnography. The publication of his diary in 1967 alerted many to the curious paradox built into his ethnographic persona – that he was both scientist (concerned with description and analysis) and fallible human being (driven by powerful subjective desires); but it was perhaps not until the 1980s that the influence of subjective elements in ethnographic work was fully addressed (Geertz 1989; Clifford and Marcus 1986; Marcus and Fischer 1986). Although recent debates surrounding the role of Malinowski in anthropology's modern evolution have certainly changed our understanding of the nature of his contribution, they have not seriously undermined the central place in the discipline that he himself always claimed, nor have they revived any serious interest in the work of another early pioneer, W. H. R. Rivers. Indeed, most anthropologists continue to follow Malinowski in denigrating or ignoring the innovative approach pursued by Rivers in anthropology (and pyschology) at the turn of the century.[5] Pat Barker, however, finds much to explore in Rivers's life. She places him at the centre of her trilogy dealing with the Great War, *Regeneration* (1991), *The Eye In The Door* (1993) and *The Ghost Road* (1995). Drawing on established sources and her own research, Barker uses her creative imagination as a novelist to pose questions about Rivers and his project which are critical to my concerns in this paper.

Barker's first book, *Regeneration*, is set in Craiglockhart, the military hospital where Rivers worked as a doctor treating patients suffering from shell-shock. Against a backdrop of the daily eruptions of war terror among his patients, with their uncontrolled outbursts of humiliating emotion and behaviour, she reveals the character of Rivers through his changing relationship with a number of key figures, including the celebrated poet Siegfried Sassoon. While this first novel is confined largely to a hospital setting, the second and third volumes, *The Eye In The Door* and *The Ghost Road*, have a much expanded social context. Moreover the processes of individual disintegration which we witness in *Regeneration* are now seen to be mirrored in society itself, as the war shatters the hold of established rules and norms. Rivers stands at the centre of this collapsing world. He is open to its fluidity, and, through his immersion in it, Barker reveals an extraordinary process of personal and intellectual transformation.[6]

Many of the startling insights, what we might call decisive moments, occur in each novel through dream, hallucination and memory. Here Barker draws inspiration from Rivers's late writing, *Instinct and the Unconscious* (1920) and *Conflict and Dream* (1923). Both volumes were published in the period between the end of the war and Rivers's death. In his exploration of areas lying between the unconscious and the conscious, past and present, Rivers made his own mental processes an integral part of his investigation. The visual capacities of the mind interested Rivers a great deal, not just because they were so pronounced among his shell-shock patients, but because he was acutely aware of his own inability, under normal circumstances, to visualise, that is, to generate visual images in the mind's eye (See Langham 1981:54.) He recognised that a key to the variable power of visualisation was contained in the intense affective qualities which attached to the fragments of memory. This faculty Rivers called the protopathic. It was characterised by its 'primitive', totalising qualities. In normal human development, the protopathic was overlaid by a higher faculty, which Rivers called the epicritic, the capacity for rational, discriminating judgement. But Rivers's understanding of the hierarchical relationship between these two faculties was fundamentally undermined by the war. He came to understand that normal sensibility required not the suppression of the protopathic by the epicritic, but their creative integration:

Drifting between sleep and waking, Rivers remembered the smells of oil and copra, the cacophony of snores and whistles from the sleepers crammed into the small cabin on the deck, the vibration of the engine that seemed to get into one's teeth, the strange, brilliant, ferocious southern stars. He couldn't for the life of him think what was producing this flood of nostalgia. Perhaps it was his own experience of duality that formed the link, for certainly in the years before the war he had experienced a splitting of personality as profound as any suffered by Siegfried [Sassoon]. It had been not merely a matter of living two different lives, divided between the dons of Cambridge and the missionaries and headhunters of Melanesia, but of being a different person in the two places. It was his Melanesian self he preferred, but his attempts to integrate that self into his way of life in England had produced nothing but frustration and misery. Perhaps, contrary to what was usually supposed, duality was the stable state; the attempt at integration, dangerous. (Barker 1993:235).

Barker here exposes the distinctive qualities of memory as particular scenes, recovered from the unconscious through the return of the protopathic, acquire an extraordinary clarity and intensity. But the meaning of this experience emerges only through its incorporation into Rivers's present. For the war has shattered the stability of a divided personality; the new integration which Rivers seeks, between past and present, protopathic and epicritic, psychology and anthropology, becomes inseparable from society itself.

Barker's novels offered me a way of thinking about my own fieldwork experiences from a place within the anthropological discipline. The development of such a reflexive consciousness was built upon a certain imaginative leap; but having taken it, I found myself eager to return to an exploration of more conventional sources as a means for anchoring what I had discovered through literature. Initially, developing an understanding of Rivers's importance to my own identity as an anthropologist led me to reject what I perceived to be lingering Malinowskian elements. (See Grimshaw and Hart 1993, and Grimshaw 1996.) But as Rivers discovered through his own work on the protopathic and epicritic, while duality might be a state of stability, it is the creative integration of separate elements which has the potential to generate new insight. For if Rivers laid the foundations for understanding the irruptions of the subjective within anthropological work, it was Malinowski who actually created the conditions in which they might occur. For integral to his model of fieldwork, the lone ethnographer cut free from the conventional structures of everyday life, is transformation through the decisive or visionary moment.

CONCLUSIONS

Although I may seem to have travelled a long way from New Year at Rizong, questions concerning past and present, protopathic and epicritic, fragments and wholes, linear and circular time, memory and autobiography, Rivers and Malinowski are, I believe, central to any understanding of what might constitute key moments in fieldwork experience.

Fieldwork is conventionally separated in time and space. It is conceived of as a bounded experience, rather like Rushdie's 'old photograph in a cheap frame' which is expected to supply, if you are lucky, a lifetime's academic publications. Situated in a linear conception of time, as a stage in the professional career of an anthropologist, it is the liminal moment which connects the past and future, the student neophyte and the fully fledged ethnographer. But fieldwork is also the point to which we return in an endlessly recursive movement.

Whether the past of fieldwork is conceived of as whole or fragmented – as an anchor for the self or deeply subversive of its integrity – depends, as I discovered myself, on where you are in the present. For when I was inside academic anthropology, writing a doctoral thesis, the pursuit of professional integration turned me into a hysteric. Integration meant repression. I was deeply fearful of the past of my fieldwork. I became haunted by reminiscences.

The past of my fieldwork was transformed, however, by the circumstances under which I wrote *Servants of the Buddha*. I drafted the book in the corner of a tiny Brixton room, while watching over the Caribbean writer, C. L. R. James, as he retreated into a fragile old age and eventual death. From that precarious position outside the academy, fieldwork became a place to which I could retreat. It was a sort of homeland, bounded and integrated, rooting my self in an imaginary past which contrasted sharply with the fragmentation of my present and the uncertainty of my future. But, of course, that moment of writing is itself now located in my past. And the meaning which I found in my fieldwork at that time has already changed. I attempted to stabilise meanings through the excavation of powerful emotions which attached to my memory; but they were never wholly within my control.

I began my essay with two episodes from the Ladakhi New Year. In presenting the passages, I reversed their order from the one in which they occurred in real time and in the narrative of my book. These contrasting moments from my fieldwork symbolise the two conceptions of time which have concerned me here. The ceremony in the temple at Rizong involved movement through linear time, carrying its participants into a future which contained and yet transcended past and present. The ritual performed on the mountainside above Julichang, however, involved a sort of circular movement. The past and present were intertwined and inescapable. In making my point of departure the integrative moment of the temple ceremony, rather than its bleak counterpart on the mountainside or indeed the separation motif which marked the begining and end of the book, I signalled my intention to end with the prospect of reintegration.

If the old anthropology built upon the paradigm of scientific ethnography situated fieldwork as a timeless moment between an unknowing past and a knowing future, integral to a sort of Enlightenment progress, the new subjective or reflexive anthropology is always in danger of becoming trapped between a past and a present in which the possibility of knowledge itself is uncertain. This sets up an endlessly circular motion which remains insulated from both society and history. I experienced the writing of my book as a moment of free subjectivity in contrast to the imposed objectivity of a doctoral thesis. But I now recognise that the narrative I achieved in that moment of individual creativity contained the seeds of a new engagement with anthropology as a wider social project. For, although at the time I wrote *Servants of the Buddha* I believed I was finally closing the door on my academic past, subsequent events have shown me that the book can be understood as the preparation for my return to anthropology.

Acknowledgement: I am grateful to my colleagues for their critical insight and generous support, especially Keith Hart, Mark Harris, Colin Murray, Liz Stanley, Peter Wade and Bill Watson.

NOTES

1. A fuller examination of this concept is contained in my work in progress, *The Ethnographer's Eye: Ways of Seeing in Anthropology.*
2. I find it interesting that Bill Watson's seminar series was the first and last time I was given the opportunity to discuss *Servants of the Buddha* in the context of academic anthropology.
3. Among the most useful sources on the history of Buddhism in Ladakh are Snellgrove and Skorupski (1977 and 1981).
4. Material relating to the role of women in Buddhism is scarce. Havnevik (1990) is valuable and contains a good bibliography.
5. Important exceptions include Urry (1972), Slobodin (1978), Langham (1981), Kuklick (1991) and Stocking (1983, 1992, 1996).
6. This process of transformation is also discussed by Slobodin (1978), Langham (1981) and Stocking (1996).

REFERENCES

Barker, P. (1991) *Regeneration*, London: Viking.
Barker, P. (1993) *The Eye In The Door*, London: Viking.

Barker, P. (1995) *The Ghost Road*, London: Viking.

Cartier-Bresson, H. (1952) *The Decisive Moment*, New York: Simon & Schuster.

Clifford, J. and Marcus, G. (eds) (1986) *Writing Culture: The Poetics and Politics of Ethnography*, California, CA and London: University of California Press.

Davies, T. (1992) *A Modest Pageant*, London: Faber.

DeBouzek, J. (1989) 'The "Ethnographic Surrealism" of Jean Rouch', *Visual Anthropology*, 2 (3 & 4).

Geertz, C. (1989) *Works and Lives*, Oxford: Polity Press.

Grimshaw, A. (1992) *Servants of the Buddha*, London: Open Letters.

Grimshaw, A. (1996) 'The eye in the door: anthropology, film and the exploration of interior space', in Banks, M. and Morphy, H. (eds) *Rethinking Visual Anthropology*, New Haven and London: Yale University Press.

Grimshaw, A. and Hart, K. (1993) *Anthropology and the Crisis of the Intellectuals*, Cambridge: Prickly Pear Press (Pamphlet no. 1).

Havnevik, H. (1990) *Tibetan Buddhist Nuns*, Oslo: Norwegian University Press.

Kuklick, H. (1991) *The Savage Within*, Cambridge: Cambridge University Press.

Langham, I. (1981) *The Building of British Social Anthropology*, Dordrecht: D. Reidel.

Leach, E. R. (1989) 'Tribal Ethnography: Past, Present and Future', in Tonkin, E., McDonald, S. and Chapman, M. (eds) *History and Ethnicity*, London: Routledge.

Malinowski, B. (1922) *Argonauts of the Western Pacific*, London: Routledge.

Malinowski, B. (1967) *A Diary in the Strict Sense of the Term*, London: Routledge.

Marcus, G. and Fischer, M. (eds) (1986) *Anthropology as Cultural Critique*, Chicago: University of Chicago Press.

Okely, J. (1975) 'The Self and Scientism', *Journal of the Anthropological Society of Oxford*, 6 (3).

Okely, J. and Callaway, H. (eds) (1992) *Anthropology and Autobiography*, London: Routledge.

Rapport, N. (1993) *Diverse World-Views in an English Village*, Edinburgh: Edinburgh University Press.

Rapport, N. (1994) *The Prose and the Passion*, Manchester: Manchester University Press.

Rivers, W. H. R. (1920) *Instinct and the Unconscious*, Cambridge: Cambridge University Press.

Rivers, W. H. R. (1923) *Conflict and Dream*, London: Routledge.

Rushdie, S. (1991) 'Imaginary Homelands', *Imaginary Homelands*, London: Granta Books.

Slobodin, R. (1978) *W.H.R. Rivers*, New York: Columbia University Press.

Snellgrove, D. and Skorupski, T. (1977 and 1981) *The Cultural Heritage of Ladakh*, Volumes 1 and 2, Warminster: Aris and Phillips.

Stanley, L. (1992) *The auto/biographical I*, Manchester: Manchester University Press.

Stocking, G. (1983) *Observers Observed*, Madison: University of Wisconsin Press.

Stocking, G. (1992) *The Ethnographer's Magic*, Madison: University of Wisconsin Press.

Stocking, G. (1996) *After Tylor*, London: Athlone Press.

Stoller, P. (1989) *The Taste of Ethnographic Things. The senses in anthropology*, Philadelphia: University of Pennsylvania Press.

Urry, J. (1972) 'Notes and Queries on Anthropology and the Development of Field Methods in British Anthropology, 1870–1920', *Proceedings of the Royal Anthropological Institute*.

6 A DIMINISHMENT: A DEATH IN THE FIELD (KERINCI, INDONESIA)
C. W. Watson

Mengingat Ruh

Two of the much debated issues which have preoccupied anthropologists since the 1970s have been the degree to which the personality of the anthropologist should be explicitly inserted into any published ethnographic account and how competent anthropology is, or should become, in the identification, description and analysis of emotion. In relation to the first, arguments have hinged on whether, since we know it is impossible for the anthropologists not to allow their subjectivity to influence what they choose to see and how they interpret it, we should give up all pretence that our account is objective and be explicit about our own involvement. In this respect, it is suggested, we are in fact doing little more than what those in the natural sciences do when they describe their equipment and the conditions under which their research was conducted. In anthropology, describing our methodology and the circumstances in which we carried out our fieldwork has long been an accepted requirement of our published work; the demand that we now be explicit about ourselves – the primary tool after all for the information gathering – and the autobiographical features of our personalities which have affected our research – from intellectual influences to emotional predispositions – seems a natural extension of the earlier requirement. If nothing else, it is a courtesy which we owe our readers, so that they know 'where we are coming from', and are thus in a position to evaluate the reliability of the research instrument which has been so critical for the elicitation of data.

Where the debate begins is at the point of deciding how much of the conditions needs to be described and what are the relevant confessional details to provide. Central to this discussion is the issue of the representation of the 'other', and indeed one might be tempted to argue that the different positions taken on either side of the debate reflect the degrees of proximity of participants to the issue of representation. That is, those who, say, fall in the environmental-biological side of anthropology where the task is not to represent other cultures but to elicit hard data which can be arithmetically measured, feel that although their credentials as trained surveyors require demonstration, their personal autobiographies are irrelevant (but see Rosaldo 1993:184–6 on Harold Conklin). On the other hand, when on the social and cultural side of anthropology, observers are constantly making judgements and evaluating the degree of significance actions have for members of a community, those observers need to be explicit about, for example, their own gender, age, marital status and their general education orientation *inter alia*, a point strongly made by Gardner in her chapter.

Of course, the issue is never so clear-cut as this, and in thinking that it is, participants on the two sides – more disclosure, less disclosure – frequently talk past each other. The anthropologist in the field is rarely simply measuring the girth of trees, and even those who see their task as becoming absorbed in the other culture write notes for future reference. Where, however, both sides concur and where the problem can be seen at its most acute is in the recording of emotions and the study of affect, an area in which for historical reasons American anthropology, with its tradition of 'culture and personality' studies (however flawed the latter were), has been in advance of that of Britain. British social anthropology under the influence of Radcliffe-Brown, taking his lead from Durkheimian sociology, deliberately eschewed the study of emotion and individual personalities, and instead insisted on the study of social facts through which one came to understand social structures. The consequence of the turn that the discipline took in Britain at that point has been not simply the avoidance of anything that hints of psychology, but a deep-rooted suspicion and hostility toward alternative anthropological traditions which seek to make personalities and emotions and their embeddedness within cultures the object of their research.

Those British anthropologists who have, then, wanted to investigate such issues have been compelled to turn to American anthropology where there has been a recent revival of interest in investigating not only cognition but affect within specific cultural contexts (Lutz and White 1986). These new foci of research can be placed in a direct line of continuity with the earlier 'culture and personality' approach, but differ substantially from it, particularly over the crucial methodological issue of how one is to measure emotion (See the useful volume edited by Shweder and LeVine (1984) for pertinent essays on the subject.) It is here above all that the issue of the use of the anthropologist herself to gauge and evaluate what is effectively occurring is crucial: how reliable an instrument are one's own emotions for identifying what is being felt by another? It is with respect to these matters above all that the reader of any ethnography must be fully assured of the ethnographer's personal as well as methodological fitness for the task if any credence is to be placed in the ethnographic account: there is a need for transparency. A convincing demonstration of the success of such an approach can be found in Jean Briggs's *Never in Anger* (1970) where she describes her own emotions in tandem with those of the Inuit family with whom she lived. It is precisely the analytical juxtaposition of the anthropologist with her informants which we find there that allows the reader dialectically to construe an understanding of another culture.

It has sometimes been argued, however, that sympathetic as these accounts are in which the anthropologist is implicated, they none the less occlude the issue of power and hierarchy (Hastrup 1992:122). Another formulation of the same point might be that these narratives too are entrapped in the Fabian (1983) dilemma in which the present of the lived event becomes distorted in the ethnographic recording which orders, categorises and evaluates according to criteria absent from the original encounter.

In the face of such criticisms it might appear that the only way to overcome the problem of hierarchy would be to empower the other: in this case by a strategy of minimal intervention in the part of the anthropologist. As the work of Oscar Lewis has brilliantly illustrated, the anthropologist is well-placed to act as a conduit for the direct expression of autobiographical voices (see Melhuus (1997) for an excellent discussion). But such a strategy is disarmingly deceptive since we know that, as with ethnographic film, the recorded voices and images have in many different and subtle ways been prefabricated by the directorial techniques imposed by the anthropologist as auteur. Far better then to be explicit and show how the anthropologist's promptings have contributed to the scenes or how his or her own feelings have pushed an interpretation in a specific direction.

In a very personal and powerful essay, Rosaldo (1993), to whom Andrew Beatty also refers in his chapter, does precisely that, showing how the grief and anger which welled up within him on the death of his wife in a tragic accident in the jungle in the Philippines allowed him access to an emotion which the Ilongot headhunters had referred to and which he had not up to that point fully understood. His own grief and rage and his reflections on those emotions not only allowed him to understand the emotions of the other but also – and here is the ethnographic point – enabled him to represent those emotions in a more articulate and apprehensible fashion.

I don't find all of Rosaldo's argument convincing, particularly in the detail, but the substance of it is surely undeniable. In the description of emotions, one has to allow for 'multiple sources of knowledge' (Kondo, cited by Rosaldo 1993:181). The detachment of the scientific observer, whether measuring change in facial expression (see Ekman, cited by Lutz and White 1986:410) or interpreting the symbolic meaning of ritual objects, by itself can never be sufficient; there has to be a way of providing for readers imaginative access to the emotional significance of events as felt by the participants. To make a start in such a task, one has to become inward with a culture, and one possible avenue here is by a confrontation of one's own emotional responses with those of the people with whom one lives.

The death of Rosaldo's wife in his case precipitated reflections not only about the nature of his own emotions but also about the anthropological discussion of death about which he makes two important points (1993:12–16). The first is that anthropology tends to focus on the bounded event because it provides for ease of analysis: it is self-contained, discrete, temporally marked-off. A description and explication of the event lies within the control of the anthropologist who alone can determine what is or is not extraneous or irrelevant to its interpretation. (One might make the same point, *mutatis mutandis*, about practical criticism in literature being much happier with poems than with extracts from novels or plays.) His second point is that the analysis of ritual, and here one supposes that he has ritual exegetes like Victor Turner in mind, often begins with the promise that the ritual itself epitomises micro-cosmically the structure of belief of a community. In both respects, anthropology, suggests Rosaldo, is failing to give due weight to other possible ways of understanding events. For one thing they are not bounded in the way suggested – the boundary making is a function of the ethnographer's categories – and one should see events such as rituals occurring within a stream of connected happenings. In his phrase they are 'intersections of busy-ness', and we must know where the roads lead to and from. Secondly, rituals, in his example, funeral rituals, are not simply or necessarily points of recapitulation nor a mode for restoration of order, nor do they necessarily represent the way in which a society comes to terms with grief. They are episodes within a much longer process of bereavement and need to be perceived as such. Ultimately, his point is that the anthropological focus on the rituals surrounding death – and he gives several examples – have failed to do justice to exactly what it is which is being experienced by those who are grieving and how that experience of grieving is not terminated by ritual but persists in different modes of articulation.

The issues which Rosaldo raises in particular, and the several debates about the 'positioning' of the anthropologist which have been alluded to above, speak very much to my own concerns in the following account of an episode in my fieldwork. As in Briggs's case, I try to describe my own emotions. It would be easy to say that the intention is to use that description as a springboard from which to understand the feelings and sympathies of the people with whom I was living, and indeed the perceptive reader will, I hope, to be able to set off what I say about myself against what I imply about others. This, however, was only a part of my intention. Since much of the point of the description is my uncertainty about how others felt, rather than be explicit in areas where I am still unsure, I have chosen instead to provide some indication of the context of the episode which occurred. Like Rosaldo I believe that it is only through the context that the reader can hope to perceive the events as they affected and continue to affect those who were caught up in them. This account, then, like the others in this book also tries to give a sense of what it is like 'being there', but perhaps different from my fellow contributors, what I am striving for here is not so much a reflective account of what I succeeded in understanding but an evocation of what I came to feel.

A lingering embarrassment still catches me here: I have been brought up in a tradition which regards feelings as suspect and the explicit discussion of them as inconsiderate and offensive; furthermore, this tradition of stoicism, if one wants to call it that, is also strongly endorsed in Kerinci, my second cultural home, and therefore I am engaged as it were in a double act of betrayal. None the less, there were strong reasons which have outweighed my scruples, not least of which is the fulfilment of a debt to those who are at the centre of my narrative. In addition I wanted to show how a significant incident of a personal nature, in this case the death of a friend and the circumstances surrounding the death, can lead an anthropologist in the field to reflect on how best he might fulfil his vocation: to make accessible the variety of human experience and thereby persuade us to take a larger view of our humanity.

In what follows I describe the death of a young girl which occurred during a period of fieldwork in Kerinci in central Sumatra in 1979. She was not very well known to me, though I knew her family well. Besides those reasons I have mentioned above, another consideration which made me hesitate before writing about Zaura's death (not her real name), was because I felt uneasy, and still feel uneasy, about what I sometimes see as exploiting the death of a friend and her family's grief, subjects perhaps best left within the privacy of the family. The writing has seemed to me a kind of betrayal of a friendship. What finally prompted me, however, to try to write was the conviction that if, as I have often claimed to friends, students and colleagues, the purpose of anthropology is to render as sharply as possible that understanding and experience, to which the anthropologist in the field has had a uniquely privileged access, for the benefit of those who desire to know what it is that women and men everywhere live by, then the account becomes not a betrayal but a celebration of friendship. True, it must remain a celebration which can only be partially understood by those who are at the centre of the narrative, who lack the entry into that special kind of discursive literary framing of knowledge which I, and the reader for whom this is prepared, take for granted, but the writing of it at least sets it down in potential, so that the celebration is always available for future realisations, perhaps by the heirs of those it describes. For those who can now immediately apprehend the celebration, this sharing of experience brings with it an enhancement of the quality of life of us all: surely the object of all anthropologists, and one which I know those in the narrative would understand and applaud.

THE CONTEXT

In 1978 I went back with my wife and two daughters to Kerinci in central Sumatra to conduct the fieldwork for my doctoral research. I say went back, because I had been there before. My wife is from Kerinci and we had been there together for the first time in 1972 and had made two further trips there during the period I was teaching in Malaysia in 1973–76. I was, therefore, relatively familiar with the area, and of course the fact that my wife was born and brought up there and her family still resided there, put me in a very different position from many anthropologists. It was not quite doing anthropology at home, but then again it was not the usual cold-bath immersion of the anthropologist's first major entry into fieldwork. There is much that I could say about the peculiarities of my position, which had its advantages and disadvantages in terms of either facilitating or inhibiting access to types of information and understanding which I sought, as well as in terms of the way in which I was expected to behave, but that is not what I want to talk about here. I do, however, need to put into an intellectual context what it was that I was hoping to learn in that time of fieldwork.

There were in fact two separate strands to my doctoral research proposal, which I had plaited together as best I could: the first was a desire to unravel the complexities of what was an unusual set of principles of kinship organisation. The second was a wish to locate Kerinci within some sort of development model, that is, tracing the socioeconomic changes which had occurred in the region over a period of 100 years or so, and linking that diachronic approach to a detailed analysis of what was happening at the present. The first of those pursuits had been a long-term ambition. Even before I had considered becoming an anthropologist, from the time when I had first got to know students from Kerinci who were studying in Bandung, I had been intrigued by their descriptions of a system which put such stress on the property rights of women; that initial curiosity had grown as I observed how much of what was said and done even so far from Kerinci as Bandung was being determined by a morality of kinship. It was, then, a slightly naive intellectual wondering which had led me into anthropology. In the years between first hearing about Kerinci and finally working there, that curiosity had become informed by a growing knowledge of anthropology which on the one hand had the enabling effect of providing me with the intellectual equipment to set about my task of discovery more systematically and with more assurance. But, conversely, the knowledge that the matrilineal emphasis of the Kerinci was not so unique and rare as I had imagined, and that there were established templates or analytical frameworks into which the Kerinci could be accommodated, detracted considerably from the wonder and mystery of it all – extinguishing the dying traces of what experienced anthropologists might regard as an unhealthy romanticism.

The interest in development on the other hand had taken flame from the still warm embers of another youthful passion – radical socialism. Arriving in Indonesia in 1969 full of the zeal of '68, hostile to what I knew from the books had occurred in Indonesia, that is, the destruction of communism in the coup of 1965, I was time and time again forced to abandon the dogma of the left and confront the immediate question of how one judges the there-and-then attempts of a government to improve the quality of life of the people of the country. Of course, it's a naive question, and posing my dilemma in terms of the competing demands of dogma and pragmatism is naive too. Inevitably, though, I was caught up in arguments which compelled me to think much more closely about the links between the objectives and the strategies of development. This drew me into the heated polemics of the development debates of the 1970s during the course of which, while never abandoning a belief in the ethical imperative of socialism – from each according to his ability, to each according to his needs – I became less dismissive of the argument of the trickle-down effect, less sure that the Green Revolution must inevitably culminate in polarisation and rural poverty.

By the time I went up to Cambridge in 1976, then, my plans were clear. I would, for the next few years at least, embark on a systematic course of study in the social sciences, in anthropology in particular, in order to satisfy myself that I did properly understand the arguments about development. At this stage I was particularly impressed by the work of my supervisor, Alan Macfarlane, on the Gurung of Nepal, in which he had conscientiously collected a luxuriantly detailed amount of information on production and reproduction in the Himalayas. I was not so sure about the conclusions which he drew from the materials, but the painstaking assembly of information was impressive. I was equally impressed by similar work which Ben White had done in the countryside in central Java, calculating on the basis of detailed statistics of income and expenditure, again laboriously obtained, that for the Javanese family it was better to maximise their family size to enhance their life-chances rather than the reverse.

I wished to emulate their conscientious methods in Kerinci. I felt that whatever conclusions I reached, after having conducted such an exercise, I would at least be in a position to speak with authority about the socioeconomic dynamics of a peasant society. There were, I remember, one or two voices at the time who questioned if I was really doing the right thing. Adam Kuper, I recall, when I first told him in Leiden about my plans and explained how I had come into anthropology, asked whether it might not be more appropriate to use my background in literature and languages to look into ritual and cultural performances in Kerinci, to which I replied that I wanted to move away from that sort of thing and investigate socioeconomic development.

By the time I arrived with my family in Kerinci, after having spent several months working through archives and libraries in the Netherlands, I had reached what I now look back upon as a peak of preparedness. I had the history of Kerinci – at least that history as reported in the archives – at my fingertips; I had read the detailed Dutch Residents' reports on Kerinci; from my previous visits I knew about recent experiments in introducing new high-yielding varieties of rice, and I had mapped out a strategy for conducting interviews and collecting statistical information on land holdings, employment opportunities, household income and expenditure. As for an understanding of the principles of kinship, having taken to heart Leach's famous remark about kinship being another way of talking about property, I was confident I already knew a lot about the ideology as well as about terminology, and that the principles of the system would become clear once I started to investigate the allocation of rights and duties within the society. It was, then, with a certain measure of self-assurance that I arrived in Lubuk Dalam.

BEING IN THE FIELD – THE CIRCUMSTANCES OF LEARNING

It seems a common experience among anthropologists, except perhaps for the ruthlessly single-minded, that the direction their research in the field takes, although it may remain within the broad area of what was plotted out in the original research proposal, is determined by the availability of opportunities and the congeniality of the personal friendships that are struck up once in the field. Certainly this was my own case. As I deepened my acquaintance with friends and my affinal relatives I found myself increasingly less concerned to foist my interests and enquiries on them, and much readier to be carried along by the flow of what interested them, willing to respond to any contribution they might ask me to make in terms of offering an opinion or giving some information, but not, unless the occasion was opportune or relevant, initiating any discussion of my own. I occasionally worried about this, and wondered whether I should be pursuing a more active strategy of interrogation and enquiry, but for most of the time I convinced myself that through the gradual absorption of the discourse of everyday life I was learning about patterns of kinship in a more organic and spontaneous – and under some criteria, more valid – way than I would through administering schedules and conducting formal interviews. As far as the interest in kinship was concerned, then, I let myself drift with a purpose.

With respect to development issues, the disposition of circum-stances was very different. Although kinship as such never loomed large in discussions, development – *pembangunan* in Indonesian and Kerinci – was forever being raised as a topic of conversation, both at a general level of talking about the strategies of central and regional government, and at a level of more direct immediacy in relation to what was happening within Kerinci itself: the initiatives which had been undertaken to create a permanent market area, improve the state of roads, provide a better supply of electricity, build more schools or improve the quality of television picture reception. At a more immediate level still, conversations went on endlessly about the rise and fall of cash-crop prices and who had made or lost fortunes through making the decision to plant coffee or cinnamon or cloves at the strategic time. It was these issues which animated the casual conversations between friends meeting for an idle moment in the resting time between *azhar* (the mid-afternoon prayer) and *magrib* (the dusk prayer), or the longer discussions after the *isha* (evening) prayers when advice would be sought about the proper way to plant clove seedlings or how to prune coffee bushes. Very often the general discussions about farming would be interrupted by very specific anecdotes relating to recent news: a metre of bark had been stripped off a large clove tree of a wealthy farmer whose arrogant manner had offended many; a farmer from Kumun had heard on the radio which he had taken up into his orchards in the hills that there had been a frost in Brazil and the coffee futures prices were rocketing – he'd sold his coffee to a Chinese buyer and made a handsome sum; a bear had found its way into one of the upland huts and gorged on the bananas which had been stored there.

As I listened to what was said, recording some of it – alas, as I now see looking back at my notebooks, not enough – I gradually began to acquire a sense of things, by which I mean I began to feel that I was moving to a better understanding of how the people of that community ordered things in their lives, what priorities they attached to their activities and to their relationships with others. And it was at this point that I wanted – without ever consciously putting it to myself in these terms – to put my understanding to the test, by more actively contributing in the discussions. I was, however, hampered by one major drawback, my ignorance of most of the activities which formed the staple of their conversation. I knew nothing about farming or marketing. With regard to the latter I could and did inform myself by following the chain of buying and selling from the producer located in the hills above Lake Kerinci to the exporter in the warehouses on the shores of the Indian Ocean in Padang. But as for farming, if I was ever to claim any authority to speak I would have to learn, as all anthropologists are recommended to do, by direct participation, that is, by farming myself.

Here, fortuitous personal circumstances came to my aid. My wife had, according to the matrilineal principles of inheritance of property, rights of access to land, both wet-rice land (*sawah*) and upland gardens (*ladang*), and, indeed, as her husband I was expected within the traditional scheme of things to work the land, above all the *sawah*. Availing myself of this duty-privilege I was, consequently, able to reply to the several enquiries of a traditional kind that were directed to me that, yes, this year I was 'going down to the *sawah*'.

Of course, it wasn't quite as easy as that; there was a lot that I had to learn about growing rice. One of the first lessons I learned was that it was not something that one usually did by oneself. There were various activities one could do on one's own, particularly if one's holding was small, such as preparing the seed-beds and sowing the germinated seeds, but the major operations – the preparation of the soil, the transplanting, the weeding and, later, the harvesting – were usually done in gangs, sometimes gangs of cooperative farmers who would work on each other's fields in rotation, sometimes gangs of hired farm labourers. It was while I was planning the organisation of these various farming activities and puzzling over how I should recruit the labour for the various tasks that I got to know Zaura and her family.

THE DEVELOPMENT OF A RELATIONSHIP

My mother-in-law every year regularly employs hired labourers to help her in the rice-growing season. The choice is always open to her to rent out her rice fields: there is a variety of complicated renting arrangements which can be contracted which I have described elsewhere (Watson 1992). She, however, always likes to retain control and responsibility for some of these rice fields each season, so that she, too, can respond in the traditional fashion when asked, that she is planting rice this year (*ikut ke sawah*).

Hired labour is, however, in short supply in Lubuk Dalam. The reason for this is that the village lies close to the central town area and most of the young men in the village prefer to seek employment opportunities in the town as casual labourers, semi-skilled artisans or in the small transport and services sector. The returns on one's labour are, at least in the short-term, much higher than in farming, where much of the return on labour has to be deferred.

A consequence of this is that over the last 20 years or so, my mother-in-law has had to look beyond the immediate vicinity of Lubuk Dalam for help. In particular she has established a relationship with a family in Rawang, a densely populated cluster of villages, about a mile downhill from Lubuk Dalam. There is a general shortage of land in Rawang, not only *sawah* but the more lucrative cash-crop growing land of the hills, since Rawang lies in the centre of the valley, and the upland has traditionally been allocated to the villages at the foothills. The relative poverty of Rawang, then, had led many to migrate out of the area, some as far as the Malay peninsula. For those who have stayed on and whose only skill is working the land and engaging in the trades associated with farmwork as minor craftsmen or petty rice traders, the only option is to seek work as farm labourers.

My mother-in-law's relationship with the family in Rawang has deepened and broadened over the years. At first, it was only the menfolk that came up to the house, and were given their instructions for the day. (Depending on what time of year it was this might mean starting the preparation of the land for planting, or transplanting the rice, weeding, checking the water supply, spreading the fertiliser.) Gradually, however, it became a family affair as the women got to know each other. My mother-in-law might go down to the fields where the women from Rawang would be working. She would sometimes take the midday food, or when it was harvest time, she and one or two of her daughters might go down to help with the threshing and winnowing. Afterwards everyone would go up to the house at the end of the day for coffee and bananas, a cigarette or a chew of betel, relaxed, joking, enjoying each other's company, or at least so it seemed to me as an observer.

The key figure in all these transactions, all this coming and going between Lubuk Dalam and Rawang, was a man we all called Wo, elder brother, not because he was old – I suppose he must be about my age – but because that was what they called him in Rawang. He was a very quiet man, conscientious and hard-working, according to my mother-in-law. Besides working as an agricultural labourer, he was also trying to make a bit of money by buying and selling rice – buying the paddy from the farmer, taking it down to the rice mill in huge sacks slung on to his bicycle, getting the rice milled and then selling it off to the rice traders at the market – it was an arduous job, but there was a guaranteed small profit at the end of the day.

In fact it was in relation to his rice trading that I first got talking to him properly. Before that we had acknowledged each other's presence but only just perceptibly, because we were both embarrassed. Our small daughter Dewi had struck up a friendship with him at the back of the house, in the kitchen, where I could go but where I always felt I was in the way; and somewhere in one of our boxes there's a black-and-white picture of him carrying Dewi, Wo looking deadly earnest and Dewi with her two-year-old's mischievous smile. I'd been glad to see the friendship, both because it was good in itself and because I had hoped it would allow Wo and me to get to know each other, but nothing seemed to come of it. We were both still too shy of each other, myself not wanting to be intrusive, and not really understanding the dynamics of the relationship with my mother-in-law, and him, I think, feeling that we had little in common, and having no desire to appear to be ingratiating himself.

When I embarked on the long period of fieldwork, however, the marketing of rice was one of those things I wanted to know about, and so it was natural that I should turn to him for information and he was quite happy to answer my questions: the prices of various types of paddy at the rice field or at the farmhouse gate, the difference between 'wet' and 'dry' paddy, the cost of milling, the various degrees of rice polishing, seasonal variation of prices, etc. We got on reasonably well, although Wo was not, and is not, over-talkative. I tried to make jokes, but he either didn't think they were funny, or didn't understand I was trying to be humorous. The only time he was emphatic was when I made some absurd statement, or showed by what I said that I hadn't understood him properly. Then, looking almost shocked, he would say '*Ndaah*', drawing out the flat second syllable, 'No (You've got it wrong)'. But on the whole we understood each other after a fashion, and we shared the sentiment of both being slightly in awe of our common benefactor, my mother-in-law.

It was natural, then, that when my wife and I wanted some help to work our rice fields, it was Wo to whom we should look. He in his turn contacted his relatives and friends at the appropriate times, so there was always enough labour on hand to do the separate tasks, the breaking-up and ploughing of the hard earth, the making and repairing of the bunds, the preparation of the seed-bed, the puddling of the wet soil to make a tilth, then the transplanting, weeding, scaring off the birds and finally the harvesting, all in the space of about four and a half months.

For me it was an exhilarating experience, watching and participating. I would go regularly to the rice fields, sometimes just to see how the rice was growing, sometimes to join Wo and his friends for lunch. I learned as much as I could of the various techniques and amused myself and them by making a fist out of certain tasks, the transplanting for example, while not being too bad at others, wielding the mattock or spraying the crop. It was all good fun and we laughed together a lot. But I was under no illusion and I knew that I was being indulged and my hobby farming wasn't taken very seriously. I took a number of photographs, and when, as we occasionally do, the family looks again, often with a sense of wry amusement, at that pictorial record of our experience, we recall the time as a series of very happy occasions.

We got to know Wo's family, or at least his wife's family, quite well during those months. It was to be expected that it would be his wife's family we got to know rather than Wo's natal family, because the kinship system is highly matrifocal. When a man marries, he surrenders many of his ties with his natal family, and becomes more and more committed to his wife's interests.

Post-marital residence is often uxorilocal, but the tie with the wife's family is more than one of residence. The husband becomes thoroughly incorporated into the kinship network and friendships of his wife's family, and much of his time and energy is spent with her kin in joint enterprises, working together with them in the fields, helping to make arrangements for the numerous ritual ceremonies held throughout the year, or simply attending to everyday domestic chores in the household. Not that the husband may not have his own circle of friends, independent of his wife's family, but his time is largely taken up with the family and, increasingly with the years and as children are born, he is associated with it, rather than with his family of origin. All this I learned only slowly through observation of Wo and others, and indeed from seeing what was expected of me.

It was natural, then, as I say, that we became more and more familiar with Wo's in-laws: his father-in-law who was a *pandai besi*, a smith who made and repaired his own farm tools; his mother-in-law who organised things; his outspoken wife, Dahlia, her face scarred by smallpox and her younger unmarried sisters, Su who had a harelip and whom I couldn't understand but who always smiled, and Zaura, the pretty sister, who giggled whenever she looked in my direction, Zaura, whose death it is I am recording here.

DULCE RIDENTEM

I never really got to know her, never ate with her or exchanged more than a few words. I watched her often, she and sisters and friends, as they got on with their tasks, caked in the mud of the paddy-fields, weeding out the grass and talking and joking all the while, and at the end of the day going to the washing pond and changing into their good clothes to walk home to Rawang. I would have liked to join in their talk, but I didn't know what to say which wouldn't have embarrassed them and me. Occasionally, when I went to the house in Rawang, they would serve me tea and bananas or cakes, but then would retire to the back, not interested in what I had to say, and leaving the difficulty of conversation to their father and their brothers.

Zaura's death was sudden and unexpected. I remember most of what happened well enough, but where the details have been forgotten my diary of the time reminds me of the events as they happened day by day. It was while we were waiting for the paddy to ripen for a few more days before harvesting, that Wo came round and told us that Zaura was very ill and the family were worried. She couldn't seem to catch her breath and didn't have any energy. She couldn't move and the family was anxious. The news came to me second-hand: it had been related in the kitchen at the back, and was then passed on to me. I came and asked a few questions, but it wasn't clear to me whether this was simply news that was being passed on, or whether some sort of response from us was expected. I said that Zaura should be brought to see the doctor in Lubuk Dalam immediately, and that I would undertake whatever the expenses were. I knew the doctor quite well, and was sure he would do what he could.

A day passed and there was no news. I wondered whether the problem had cleared up or whether the family had tried some other expedient. Then in the afternoon we received a message that the family were bringing Zaura to see the doctor, and could I make the arrangements. I wrote a note and the doctor said he would be happy to see her. An hour or so later the family arrived. The men had come on their bicycles, and the women had come with Zaura in a horse-drawn buggy.

It was a confusing, disorienting experience for us all. The women were dressed in visiting clothes, in fine sarongs and blouses with brightly coloured diaphanous shawls – even Zaura was dressed like this. I was shocked. The brightness of her clothes was in disturbing contrast with her appearance: she sat with her back to the wall, very pale and expressionless, oblivious, it seemed, to what was happening around her, clearly very ill. There was some discussion in our house, and then they went off to the doctor. I didn't go with them, since I was having trouble with my leg, but my wife accompanied them to give the necessary explanations to the doctor.

An hour or so later she was back. The doctor wanted Zaura to go into hospital immediately, but the family were uneasy. Hospitals were places where you went to die, in their scheme of things. At this point, however, I became assertive: if the doctor said she should go to hospital she must; again I said I would bear the expenses. Looking back now at that time I marvel at this assertiveness of mine: was I role-playing, and if so what role: the patron telling his clients how they should behave?

The person of education riding roughshod over the ignorance and prejudice of the uneducated? There was perhaps something of both. We all talked it over, and my confidence that going to the hospital was the right course of action seemed to persuade everyone. The question of how much everything was going to cost was an issue, but relative to the general fear of hospitals and the people who ran them, this was of minor concern.

So Zaura went off to hospital and one of two of the women of the family stayed at her bedside. She was put on a drip and given medication and things seemed to be going well. We went and visited her in hospital and there were now smiles on everyone's faces. I began to take a pride in the fact that it had been largely me who had been instrumental in getting things done, a pride which on reflection I might now regard as something near smugness but which at the time seemed to arise merely from a quiet inward satisfaction at having done the right thing.

On the third day we learned that Zaura had been taken home, apparently having recovered much if not all of her strength. We were all pleased at this outcome and the passing of the crisis. The next day, however, we heard there had been a relapse, and Zaura's condition had rapidly deteriorated. My wife and I immediately went off to Rawang – my leg had healed by now – and were again shocked by what we saw. In the big front room of the house there was a crowd of people sitting on the mat, their faces drawn and anxious, talking animatedly in low voices. Zaura sat in a corner of the room, much worse than I had seen her before, gasping for breath, listless, horrifyingly different from the giggling girl in the paddy-field whose image kept coming back to me.

I went straight back to the hospital to find the doctor who had been treating her to ask what had gone wrong. It transpired that there was a problem with Zaura's heart, and it was this which was causing the breathing difficulties. In fact the doctor had wanted Zaura to stay in hospital and it was against her advice that they had taken her away. It was imperative, she said, that Zaura be brought back. By this time I was feeling very anxious and much less sure of myself. Back at Rawang I said they must take Zaura to the hospital again, and they, too, now very worried and bewildered with no clear ideas of their own about what should be done, agreed. While they made preparations my wife and I went back to Lubuk Dalam.

At two-thirty in the morning there was a knocking at the door of the house. I knew what it must mean. I lay in bed and heard the drawing back of bolts and the whisperings at the door. My wife went out to investigate and came back a few minutes later.

'*Zaura sudah dahulu pergi*', she said, 'Zaura's gone on before.' She'd died an hour earlier. Dahlia had been sitting with her and she'd stopped breathing and Dahlia had been unable to rouse her. They were taking her back to Rawang now for burial; they hadn't been able to get the use of the ambulance and so they were carrying her back on a litter to bury her near the house. Everyone seemed very calm, almost relieved. The anxiety had been acute in the period leading up to her death, when the procedures and the action to be taken were unclear and they were unsure. Now that Zaura was dead, they were once more in command of themselves: there were established and helpful ways of dealing with death. On the other hand, my position was reversed: leading up to the death I had been sure of the measures to be taken and the procedures to follow in order to deal with crisis; now Zaura was dead despite my actions, and I no longer had any solutions. It wasn't, however, simply a feeling of personal frustration and help-lessness; it was also a knowledge that in this new circumstance I, unlike the others there, was ignorant not only of the tasks that had to be carried out, the formalities to be observed and the rituals that needed to be conducted, but, more worryingly, I was without any personal and cultural orientation that could give proper shape to my attitudes and emotions. What eventually forced its way through, through that unsorted assemblage of experiences which constituted my personal identity, was a feeling of overwhelming sorrow, for the family, for Zaura, for myself, for the limitations to human life which circumscribe us all; sorrow for what might have been and could never be, sorrow for what had been and was now gone for ever.

The funeral was held later that morning, after the corpse had been washed and dressed. I felt utterly desolate, but I was expected to attend. I was worried about how I would control my tears: I didn't want my visible emotion to embarrass them. They were waiting for me at the house. Zaura's body was laid out with a sheet covering her. They drew it back for me to look at her face, very peaceful in death, with flecks of powder on her lips and cheeks. I looked on her for a minute or so in silence, and then they drew the cover over her face again and proceeded with the rituals of burial.

Two days later the family from Rawang came to the house to pay us a formal visit. It was the occasion on which they wanted to thank us all for the support we had given, and to ask what they owed us for the various expenses which had been incurred. I said there was no debt, that this was what one did for family and friends. Again they thanked us, and said we were indeed family – they used the intimate and critical Kerinci phrase '*mengaku dusanak*', 'to recognise as kin' – and we would always be welcome in Rawang. It was a sombre occasion with the memories of the last week very recent and little to lighten the atmosphere.

Five days later we all began to harvest the rice which Zaura had helped to plant.

RETROSPECT

The motivation which is common to all of us who want to become anthropologists is the desire to make sense of examples of human experience which at first glance seem to us to be different, unusual and alien. We begin with the assumption that the experience is comprehensible, that there is an underlying logic to the way people behave and the decisions they take, and that our first task is to accumulate the information that will enable us to understand the logic. Most anthropologists, if they are any good at all at their jobs, eventually manage to achieve this understanding. For them, the unfamiliar becomes familiar, assimilable within their way of looking at the world, or at least translatable into common terms of understanding. The second task which we take on as anthropologists is much harder, and very few of us accomplish it well, that is, the rendering of our understanding into a form which will give others an opportunity to share not only our visual experience and sensory observations, but also our interpretation of what we have perceived, and our intellectual and emotional appreciation of those perceptions. It is, above all, the last task which is impossibly difficult, since the anthropologist distrusts his (I had better use 'his' here since I'm talking about myself) own emotions and regards them as idiosyncratic and improper indices of a total apprehension of the experience. My tears, my joy, my laughter spring from the very particular and individual and highly specific quality of what I have lived through and known. On the other hand, there are commonalities which I share with those for whom I write, and therefore I should be able to guide the reader through an approach which allows her to understand both the idiosyncrasy and the commonality of experience, and consequently enables her to share some of what it was like to have been there and to have known and felt what I did.

Very often, because of the difficulty of rendition as well as the fear of self-exposure and the risk of that personal emotion appearing bizarre or sentimental or absurd, anthropologists avoid the issue. There are two ways they do this. One is by self-mockery and humour, which, while retaining for the anthropologist a central place in the narrative, effectively undermines any attempt to convey the significance or the intensity of the encounter, and consequently confines anthropological explanation to the objective analysis of observed behaviour. The other way to avoid the issue is simply to refuse to recognise it as a problem, and to draw a sharp line between the account of the measurable phenomena and those other features of social life, the emotions and sentiments. In relation to the latter, the difficulty of recording and describing is seen to lie in the impossibility of the observer extracting his own emotional involvement when embarking on the task of gauging the intensity of profound emotion. British social anthropology, as it has developed, seems especially prone to adopting the latter solution, leaving the measurement of affect to the psychologist, and the translation of emotion to the novelist.

That response, however, strikes me as a shirking of professional responsibility: in taking up his vocation, the anthropologist has accepted a commitment to convey a total sense of the other society. In some ways so has a feature journalist, an essayist, a Paul Theroux, or a V. S. Naipaul, or a Colin Thubron, who do that sort of thing very well, and who, according to some anthropologists, should be left to get on with it. The difference, however, is this: the journalists don't have the knowledge of anthropologists acquired from years of training and experience. The journalistic impressions of having the knowledge is a confidence trick, a rhetorical sleight of hand: a graphic image, a well-honed anecdote, a clever phrase, a colourful metaphor, all buttressed with the implicit statement, 'I know, because I was there.' As anthropologists we know the phoniness of many of these accounts. Even when we don't know the region or the society being described, we can tell when the description doesn't sound right.

It is, then, our responsibility alone to pass on to the reader as nearly as possible that knowledge, experience and emotion that we – through the privileges of our position – have acquired: our texts must reproduce the texture and colour of the life as we have come to recognise it. Of course this is not simply a responsibility to the discipline; more pressingly, it is a responsibility to the people who befriended us, who helped us directly and indirectly, who gave us something of themselves, in order that we could sympathetically interpret their experience in a way which would bring them closer to us – just as our explanations in the field of our own origins were designed to make us comprehensible to them. It is surely this striving for a convergence of sympathies which makes anthropology the supreme humanist discipline.

The account of Zaura's death was an attempt to fulfil an obligation, to try to convey more intimately than I have done so far in my writing the quality of life of the people I know in Kerinci. Of course it's partial, a single event which is, despite my attempt to provide some context, largely detached from the wider set of circumstances in which it occurred. I said above that what I had set out to do was to understand and then write about kinship and development in Kerinci. My writings to date do indeed cover these subjects, at least ostensibly. Here, however, I wanted to show that anthropologists have further ways of conveying an understanding of other cultures: that the systematic description of institutionalised regularities and social patterns is insufficient and needs to be set against accounts, recognisably personal and yet, at the same time, for that very reason, recalling a universal trait which makes us all kin: the capacity to construct ourselves as individual subjects. I'm not confident, however, when I read over what I've written and when I recall the experiences of the time, that I have succeeded, even partially, in evoking a sense of the emotional intensity and sudden intimacy of those relations of kinship and friendship.

For me at the time this was a critical series of events: I was much affected by what happened and the way in which I subsequently perceived what was occurring in the stream of everyday life around me from then on took on a different hue, darker but also warmer. But, however significant this was for me as an individual, did it change me as an anthropologist? Yes, but not in any spectacular or readily observable manner. I continued to gather data, to conduct interviews, to note down observations in diaries and notebooks, to participate in the sacred and secular rituals of the community. What had changed, however, was my sense of what should be the purpose of anthropology, and with that came also a change in how I felt an anthropological perception of culture and society should be recorded.

Having previously been concerned with a precise and meticulous documentation of statistical details which I had felt would allow me to speak with authority about socioeconomic change in Kerinci, I now found myself preoccupied by how I might devise a way of representing the life that I saw and was now beginning to understand, by balancing the objective record with subjective awareness of it: analytical inference matched with impressionistic reflections.

Years later this still remains for me an elusive ideal, a perpetual struggle to hammer and fashion into shape anthropological descriptions which satisfy the demand that I speak to a general condition open to cross-cultural comparison, yet which at the same time correspond in their particularity to a range of experience of which we all individually have, however indirectly, personal knowledge. The accounts never quite achieve what I desire; there is at the end of the writing always a residue of frustration, a feeling of not quite hitting the mark, but the memory of Zaura and other dead friends, the friendships with others still alive and the vivid recollection of important encounters in my life propel me ineluctably to try again the heat of the anthropologist's forge.

REFERENCES

Briggs, Jean L. (1970) *Never in Anger. Portrait of an Eskimo Family*, Cambridge, MA and London: Harvard University Press.

Fabian, J. (1983) *Time and the other. How Anthropology Makes Its Object*, New York: Columbia University Press.

Hastrup, Kirsten (1992) 'Writing ethnography. State of the Art', in Okely, Judith and Callaway, Helen (eds) *Anthropology and Autobiography*, London: Routledge, pp. 116–33.

Lutz, Catherine and White, Geoffrey M. (1986) 'The Anthropology of Emotions', in Siegel, Bernard J., Beals, Alan R. and Tyler, Stephen A. (eds.) *Annual Review of Anthropology*, 15: 405–36.

Melhuus, Marit (1997) 'Exploring the work of a compassionate ethnographer. The case of Oscar Lewis', *Social Anthropology*, 5, 1 (February): 35–54.

Rosaldo, Renato (1993) *Culture and Truth. The Remaking of Social Analysis*, London: Routledge.

Shweder, Richard A. and LeVine, Robert A. (eds) (1984) *Culture theory. Essays on Mind, Self and Emotion*, Cambridge: Cambridge University Press.

Watson, C. W. (1992) *Kinship, Property and Inheritance in Kerinci, Central Sumatra*, Canterbury, England: Centre for Social Anthropology and Computing and the Centre of South-East Asian Studies, University of Kent at Canterbury.

INDEX